BOUNCE BACK, BOUNCE BETTER

Bounce Back, Bounce Better

The Transformative Mindset for Resilience, Power, and Abundance

Brad Scot Johnson

©2024 All Rights Reserved. No portion of this book may be reproduced, stored in a retrieval system, or transmitted in any form or by any means—electronic, mechanical, photocopy, recording, scanning, or other—except for brief quotations in critical reviews or articles without the prior permission of the author.

Published by Game Changer Publishing

Paperback ISBN: 978-1-964811-50-5
Hardcover ISBN: 978-1-964811-51-2
Digital ISBN: 978-1-964811-52-9

www.GameChangerPublishing.com

DEDICATION

For my very heart-beat, my wife Karen, for our children and grandchildren, for the people I have had the privilege to serve through the years, and for each person who walked in when others were walking out… I am alive, writing, and living because of each one of you.

Read This First

THANK YOU for reading my book! To express my gratitude,
I would like to give you a free BONUS!

Scan the QR Code:

Bounce Back, Bounce Better

The Transformative Mindset for Resilience,
Power, and Abundance

Brad Scot Johnson

www.GameChangerPublishing.com

―――――――――――◆―――――――――――

Brad Scot Johnson's comeback is a resounding story of faith and grit. He is my friend and Pastor. I'm honored to have had a part in this story. After reading *Bounce Back, Bounce Better*, you will believe that ANYTHING is possible for your future.

–Kris Jenner

―――――――――――◆―――――――――――

Table of Contents

Introduction 1

Chapter 1 – Dark Days and Nights 3

Chapter 2 – "Not-Enough-Ness," Secrets, and Getting Back Up 13

Chapter 3 – A Strength That Left Me Weak 21

Chapter 4 – What to Live For 35

Chapter 5 – Belief in the Middle of Unbelief 55

Chapter 6 – Love is in the Air 61

Chapter 7 – Old Dogs and New Tricks 69

Chapter 8 – Dreaming a New Dream 77

Chapter 9 – Mindset Matters 85

Chapter 10 – The Death of How I Thought It Would Be 93

Chapter 11 – The Death of Who Others Expected Me to Be 109

Chapter 12 – The Death of Fear 123

Chapter 13 – The Death of Living Small 137

Chapter 14 – A Custody Battle 149

Bonus Section: A Resource to Access the Key Principles Found Throughout the Book 155

Introduction

On a hot Southern California summer day in 2009, Bruce Jenner walked into the Starbucks where I worked as a 50-year-old, part-time, $8.25-per-hour barista. I was awestruck. Bruce, the 1976 Olympic Decathlon Gold Medalist, was a sports hero to me. You can imagine my shock when he told me that his wife, Kris Jenner, had sent him to find me and that they had been looking for me for six months. He said she had an offer she wanted to make me. At that moment, *everything changed*!

You have no doubt seen a tennis ball bounce, but have you ever seen it in super-slow motion? The ball descends toward the ground, and upon impact (remember, you're seeing this in super slow motion), the bottom of the ball flattens. If we freeze the frame right there, it appears the ball is deflated, misshapen, and ruined.

But, if you press the play button and watch, the flat, seemingly deflated ball is really gathering energy from within, using the floor like a launch pad. Before your very eyes, you witness incredible power propelling the ball back up. You just witnessed *Bounce*.

Anyone watching my life the previous two years would have described events like the ball hurtling to earth. They could describe a forceful collision at the bottom and would have concluded my warped and misshaped life was ruined. *I* had concluded that my life was ruined. Little did I know when I started my shift that day at Starbucks that a spiritual kinetic energy was building in and around me that would propel my life up again and that I would literally experience *Bounce*.

According to neuroscience, when we bounce, it's not just back. We bounce back better! Because of neuroplasticity—the ability of the mind to adapt, change, and create new neural pathways—any bounce back is going to be better. There will be better understanding, better experience, better wisdom, and, consequently, a better future. This mindset is transformational.

CHAPTER ONE

Dark Days and Nights

Sir Isaac Newton's third law of physics states: For every action, there is an equal and opposite reaction. To envision that principle, imagine a tennis ball thrown to the ground with force. What happens? *Bounce* happens!

When a tennis ball hits the ground, several processes from the fields of physics and energy contribute to the bounce back. Please stay with me for this brief bit of science. Initially, upon impact, the ball undergoes deformation due to the compressive forces exerted by the ground. This deformation involves the re-arrangement of molecular structures within the rubber material, causing potential energy to be stored. (I love the term "potential energy.")

As the ball continues to compress, the stored potential energy is converted into kinetic energy, resulting in the ball's rebound. The elastic properties of the rubber play a crucial role in this process. The rubber material possesses a high elasticity, allowing it to regain its original shape and release the stored energy rapidly.

When the ball rebounds, the stored potential energy is transformed into kinetic energy, propelling the ball upwards. The elastic nature of the rubber enables the ball to regain its shape and efficiently transfer the energy of deformation into the rebound motion.

It is the powerful combination of potential and kinetic energy that facilitates the collapse and subsequent bounce back of a tennis ball when it hits the ground.

And just like a tennis ball, people are filled with "potential" energy. Watch as their lives compress, fall apart, or become difficult. Just like the tennis ball, people have this amazing ability to bounce back and even bounce better.

When things go wrong, people can feel sad, frustrated, or scared. Perhaps you have. It's like the ball getting deformed when it hits the ground. But just as the tennis ball can store energy when it deforms, people can gather strength and determination from their experiences.

This stored energy is from the lessons they learn and the support they receive from others. It helps them find the courage and resilience to overcome challenges. They can use this energy to make positive changes and find new ways to improve their lives.

When people bounce back, it means they are finding their way again. The bounce is their tangible expression of grit and resilience. Those who *Bounce* start rebuilding their lives, finding happiness, and achieving their goals. Often, they even *Bounce Better,* reaching new heights they never thought possible.

So, just like a tennis ball bounces after hitting the ground, people have the amazing ability to overcome tough times, gather strength, and come back stronger and more successful in the end. This is why we see men and women who had a setback follow it with a comeback; they follow a breakdown with a breakthrough.

Why? They learned to *Bounce!* And where is the energy for this bounce stored?—In the mind. The secret power to bounce is this transformative mindset.

Conversely, why do some people stay down when they get knocked down? They hit the ground hard, but they don't bounce back, they don't bounce higher, and they don't bounce better. They haven't learned to *Bounce;* they don't yet have the mindset for it. The good news is they can develop this power source.

For a couple of millennia, the sage wisdom has held true that a person—any person—can get knocked down seven times and bounce up eight.

Bounce!

It's not just a law of physics or the stuff of ancient truth, and it's not just something only a chosen few can experience. *Bounce* is possible for you: when your business crumbles, you take a gut punch from your marriage, your heart breaks as a parent, or you find yourself drowning in debt. *Bounce* is for every person who has ever had a dream that died or a part of your life hit the ground hard.

The story—*your* story—doesn't have to end with a broken dream or with a future that lacks happiness, empowerment, and success. You can learn to bounce, and not just once. It's a way to live life victoriously and confidently, knowing that whatever knocks you down does not possess the power to keep you down.

How do I know? Because I'm living proof.

Let me take you back to the night of my third suicide attempt.

*****

"BRAD!"

Someone shouted my name from far away. Then, again, "BRAD!" This time, the shout was louder and closer. The firm grip of harsh hands shook my body. My eyes fluttered open, and I tried to focus on a face, a blurred cloud. "Brad!" I felt the warmth of someone's breath.

I nodded in response. Looking around me, I tried to make sense of where I was and what was happening.

It started coming back to me.

On one side of my bed were two paramedics with pads poised to shock my chest. The person to my right, close to my face, the one who had shouted my name, was a sheriff department's deputy. I noted the badge and name tag on his uniform.

I pieced together the details.

"Brad," the deputy said, now in a normal tone. "Do you know where you are?"

I nodded.

"Can you tell me what happened?"

"I took too many pills," was my weak answer.

"I have your brother on the phone." He handed me his cell. Why was my brother on the phone? How did they even know who my brother was?

"Scrud?" I used the nickname we had for each other from childhood. My voice was raspy and dry.

"Scrud," he said with reassurance. "Answer their questions, tell them only that this was an accident, let them check you out, then call me back. I convinced them I would assume responsibility for you. And I assured them this was not intentional."

"How did they know to call you?" I asked, my mind still walking through a haze of pills and alcohol.

"You called me," he responded. "To tell me goodbye. I called 911."

Every piece clicked into place. I remembered taking the pills, a handful of them—sleeping pills, antidepressants, Xanax—all swallowed down with copious amounts of alcohol. This was my third attempt at suicide, and I was determined I would get it right this time. And once again, I failed. While these strangers hovered in my bedroom, I remember thinking, *I'm even a failure at failure.*

Story of my life—at least in the two years that preceded this moment.

What I didn't learn until later is that my brother, a Suicide Hotline Counselor and professional social worker with the Los Angeles County Department of Mental Health, convinced law enforcement not to hold me under a "Fifty-One Fifty," which is a mental health hold in a psychiatric lockdown. It is an imposed 72-hour confinement for those deemed a threat to themselves or others.

My brother was well known in settings like these, having won awards with Los Angeles County for programs he designed and implemented. He knew the ropes, what to say and what not to say. He convinced the officers that I had accidentally overdosed and meant no real harm to myself. His lie kept me free from mental health jail. He told me afterward he would bust my chops if I didn't heed his every dictate for the next steps he told me to take, including an immediate appointment with the therapist I had been seeing. I agreed to that.

In the moment, confusion rattled around my heart and swam circles in my mind. I couldn't conceive how I was still alive. *I didn't want to be.* Further, and perhaps much more significantly, I didn't believe I deserved to be. But here I was, still a failure, and now with another failed attempt to end the problem.

I was the problem.

Those who commit suicide are often considered selfish because of the immeasurable pain they leave in their wake. Many throw the accusation: *They escape their pain while leaving the rest of us to reel in the carnage of their terrible decision.*

I understand that thinking, but I do want to offer another perspective. And as someone who now has counseled dozens and dozens who walked a similar "suicide-attempt-path" like mine, there is another reason people commit suicide that *feels* anything but selfish.

For me, I had done great harm. I had broken my family, shattered a church congregation, and abandoned my faith. I abused alcohol and drugs and was unfaithful to my wife. I shudder to consider the damage I did to my family, the church I loved, and the faith of people who trusted me to be a good example. I was the problem for so many people. And each time I encountered someone harmed by my destructive choices, I saw fresh pain spread across their faces, which reminded me that I was the problem.

In my flawed and unhealthy thinking, my rationale for suicide was logical. *If I am the problem, I will solve the problem.* I ran laps around this thought process: *Everyone will be better off and be able to move past the harm I caused if I am no longer around to remind them of that damage.*

So many have graciously told me, through the many years since, how thankful they are that I failed in my three attempts and lived to tell my story. I, too, am thankful. I came dangerously close to achieving a permanent solution to a temporary problem. At the time, I saw no other way.

Laying there in the bed, aware that someone had covered up my nakedness and taken all my pill bottles, I wasn't as clear-headed as I would later become. The prevailing thought that inched its way forward in my mind that night was: *Why would God want me to live?* He, above all others, surely knew I was beyond His grace, beyond His reach, and had run through every last one of the second chances—and third, and fourth—that He offered.

It was confusing, after all the harm I had done to so many, after the disappointment I was, after the devastation I had caused my family.

There was still the lingering effect of drugs and alcohol in my system. I couldn't find a clear response to my question: *Why?* But I did know that *if* there was a reason for me to be alive, I needed to figure out what it was.

That question set me on a journey, and that journey, in fact, frames my story. It's really a story about many deaths within me that led me to the most incredible life I am thankful to live today.

This book is for everyone who ever saw life fall apart—for whatever reason—and saw relationships explode while their hopes and dreams imploded. You *CAN Bounce Back* and *Bounce Better.* You can move from the death of your dream to your dream flourishing. It's a book for all who wonder if they can begin again or begin for the first time to build the life they were destined to live.

What I discovered on this wild, wonderful journey to the life I now have is:

No matter who you are, what's happened to you, what you've done, how terrible it becomes, or how hopeless it seems, everyone can Bounce Back and Bounce Better.

You may not feel this hope or believe you have *Bounce* within you, but the truth is, anyone can develop a *Bounce* mindset characterized by grit and resilience.

To get to the deeper principles and to the steps for your happiness, empowerment, and success, we are going to need to roll the tape back even further.

BORN TO BE A STAR

As a child raised in North America, I enjoyed the universally loved experience of shaping and molding Play-Doh. The malleable, clay-like goo felt so good as I squeezed it between my little fingers. I had tools to shape and sculpt almost anything—a dog, a flat flower on a page, a person's face.

My favorite Play-Doh apparatus was a shaping machine. I would put a large wad of dough into one end of a chute and press down hard on the attached lever, and out the other end would come the shape I wanted. Here's why: Across the exit hole, I would slide a flat piece of hard plastic with pre-cut openings. On that piece was an opening shaped like a square. Slide it further across the opening, and there was a circle or a triangle. My favorite shape was a star. I would push the lever, and out would come a long rope shaped like a star.

That image stuck with me through the decades. I now see it as a metaphor for my early years. From my earliest memories, others around me pushed and shaped my natural and God-given abilities through the star hole. In their mind, that was my destiny. And though I often felt uncomfortable with the pressure, I learned early to hide self-doubt and my deep insecurities.

What good would it do to protest? What if I spoke up and said, "But I don't want to be a star," which really meant, *I don't want to speak in public. I don't want to be the lead in the school play. I don't want to be president of the after-school Bible club.* Would it have changed anything? Unlikely.

For in addition to the Play-Doh pressure I felt pushing me toward leadership and public speaking and a not-so-subtle push toward service in churches, I was taught to excel on the platform of athletics.

My parents' mantra for my brother and me as we engaged in sports was: *Play hurt.* Or, we often heard, "You can cry, but you better be getting up when you do because Johnsons always get up."

The lessons my brother and I learned about persistence and refusing to quit helped us both accomplish a lot and at very high levels. Being involved in sports gave me a respite from times of turmoil at home, and athletic teams helped me fit in as we moved from school to school, changing cities and friends nearly a dozen times along the way.

However, the point that can't be missed is we were also put into situations that were neither comfortable nor anything we wanted. An example of this was the pressure to be the church star, which goes all the way back to my earliest memories. Things my dad would say. Things others would say. My paternal grandmother introduced me to strangers by saying, "This is my little preacher. He's going to be something special one day." The layer of expectation lay like a concrete blanket over my spirit, repressing rather than building my confidence. I felt small and limited each time something like that was said. Tremors of fear tingled through my heart.

My grandmother and father meant their words to be uplifting. They were spoken, no doubt, from a place of love and confidence in me. But to me, it felt like the pressure of being shoved through the "star" hole in my Play-Doh set. With shoulders hunched, I forced a smile.

I never had a vote on the issue and knew intuitively and explicitly what was expected. The lever had come down, and a star had better come out at the other end.

Both of my parents have passed, as has my only brother. It's surreal to be the last person standing in my family of origin. Many parts of my upbringing were blessed. I know my parents loved me, and I do have happy memories. Other memories were super hard and dysfunctional.

Many of you can speak to your own experience of dysfunction in your childhood and how that shapes the person you become through life. Here's what it did to me: I came out of those years at home with three huge influential beliefs. (Here's a hint: So much of what we need to achieve requires "unthinking" thoughts and beliefs you have carried from childhood.) We'll get to that as we learn the winning mindset necessary to move beyond the death of a dream and how to pursue your destiny.

What I carried from my childhood were three life-shaping, life-altering beliefs:

1. "Not-enough-ness"
2. Secrets
3. Johnsons get up

Two of those nearly took my life. The third one helped me get my life back. Let me explain.

CHAPTER TWO

"Not-Enough-Ness," Secrets, and Getting Back Up

In my parents' later years, there was a calm sweetness to their dispositions and temperaments. Their last years on Earth were some of the happiest we had together. That was not always the case. As a little boy, I would cringe and hide while my parents yelled and occasionally threw things across the room to make a point. Anger flared hot and unexpectedly… a lot. I was an undersized child, and I vividly remember the feelings of smallness and powerlessness in those moments. Sitting in dark corners of the house, thin, bony knees drawn to my chest, trembling and crying, I would pray to God, who seemed not to hear; nothing changed during those years.

To be fair, when describing my childhood, the best I can do is offer up my perspective. Lest any of us ever believe ours is the only perspective by which to judge an event, let me remind you that my mother had a whole life of experiences and beliefs behind all her actions and perspectives. I don't know what her perspectives were, but for sure they were a big part of the scene.

My father had a whole life of experiences and beliefs behind all his actions and perspectives. I don't know what his perspectives were, but they were certainly a big part of the scene.

The same is true for my brother, though he and I talked frequently about our childhood through our adult years and shared very similar perspectives and memories.

So, when I write about my parents and my childhood, I have only one-fourth of what was going on. But here's my twenty-five percent.

Life at home for both my brother and I was up and down. Sometimes, it would be humming along, and then Mom and Dad would fight. One or the other would storm out and drive off, leaving us to wonder if our family would ever be whole again. I'd sit in a corner, knees drawn to my chest, my body quaking, while silent tears wet my cheeks.

My brother kept his emotions bottled. In all our lives, I only saw my brother cry twice. On the other hand, I was an open faucet. Tears rolled easily and freely. They still do. But when my tears weren't enough to stop the fights, I became a comedian. Even then, my attempts to lighten the mood or convince Mom and Dad to apologize to each other were not enough.

That's likely when my "not-enough-ness" developed. Very early on, I learned that I wasn't capable of solving the big things—like my parents' fights.

When it came to academics, bringing home a "B" on the report card was met with a smile and a statement like: "That's good, but an 'A' would have been better." I was being measured, and I didn't measure up.

On the athletic field, there was palpable disappointment if my brother or I lost. It's one thing for parents to tell you that you're a winner and that you can achieve and accomplish, but that gift is removed when there's a judgment or disappointment when you lose. I succeeded more than I lost, but the losses left the impression that I could only go so high, only win to a point, and I wasn't enough to go further.

My parents loved my brother and me. No doubt. We had many happy experiences growing up, and as I mentioned, in their later years, I had a nearly perfect relationship with Mom and Dad. But by my adult years, the idea that I wasn't quite enough was buried in my subconscious mind. Like an app that's running in the background on your phone, beliefs deeply held in the subconscious mind inform your emotions, decisions, and behaviors, draining your physical and emotional batteries faster than they should without you knowing why.

Perhaps you can relate. Do you see the influence of your upbringing and childhood in your patterns, default thoughts, or assumptions? Often, how we perceive the world, what we believe about ourselves and our potential are programmed early and deep. By becoming self-aware and mindful, we can begin to identify where these beliefs are true and where they distort the real picture. This is how we begin to root out the things that hold us back or keep us down.

Occasionally, throughout my life, I would have meaningful wins. As a high school wrestler, I was undefeated in my senior year through the regular season. I was named our high school's first winner of the Athlete of the Week award and featured in a news article about it. When I won all three matches to take the Sectional crown, I remember feeling not accomplishment or joy but surprise, almost like I didn't deserve it or that a mistake had been made. I didn't know it then, but it was an early version of "Imposter Syndrome."

I once entered a 10k race, after which a raffle was held. When the winning bib number for $2000 was announced, I looked in my hand to discover that it was me. Again, the thought flashed through my mind: *This doesn't happen to people like me.*

I didn't even know who "the people like me" were, but I knew enough to believe we don't win things like this.

When I was preparing to graduate with my master's degree, after years of study of Biblical languages, theology, philosophy, Church history, and related fields, I remember having an aching fear that I'd missed an assignment or the professor had misgraded a paper and soon someone would realize their error, and I wouldn't be awarded the diploma. I couldn't believe I deserved it. Imposter Syndrome.

The exact same experience happened when I was in line to receive my doctorate. That was an additional three years of school carrying an "A" average while working full time, being a husband and father to two small children, *and yet*... I thought they might not grant it to me.

I'm not describing the occasional momentary thought. This was a deep, visceral feeling of heightened anxiety that lingered chronically. I barely slept in the days leading up to graduation for fear it would all be taken away. Because, at the end of the day, I didn't think I deserved it. I wasn't enough.

Religion also played a part in this mindset. While I was growing up, my father was a pastor of many churches. He was raised by a single mother in a very conservative and strict religious background, as was my mother. They sincerely served God and loved the churches they served, and their beliefs were the ones handed down to them.

But they were sincerely wrong in some of their beliefs.

Their theology—the belief about God—was one of a *Policeman-God*, hovering just out of sight, waiting to catch you being bad. And make no mistake, their theology taught: You are bad. So, there were these mixed messages of "God loves you" and the competing message of "God is going to catch you, and when He does, it won't end well."

Fear, guilt, and shame were the three prominent emotions of my religious upbringing.

If you look underneath these, there is a screaming message: "You. Are. NOT. Enough."

Many of you reading these pages will recall similar feelings and thoughts from childhood. Each of those feelings and thoughts is encoded and deposited into your subconscious mind and there they remain, always running in the background, influencing your current thoughts about yourself, your beliefs, your emotions, and even your levels of accomplishment or success.

The irony of it all is that, while building this internal messaging system that reminded me daily I wasn't enough, there were the "voiced" messages from my parents of: "You can do it. We believe in you. Whatever you go after, you can achieve."

So, why were my accomplishments capped? Why did I only go so far and then wobble in my belief that I could go higher? Why would I elevate in some areas only to sink back down to previous, lower levels?

Don't miss this: You can only go as far as what you believe about yourself, what you believe you can accomplish, and what you believe you deserve.

Your logical mind often submits to your emotional mind, and it's in the emotional mind where self-doubt and "not-enough-ness" shout.

I didn't believe I deserved to go to the extent of my dreams or as far as what some said was my potential. Unworthiness bent me on the inside. I thought I had some potential, I had goals that were bold and a few dreams that were significant. But while my logical mind could think of them, the emotions I held about myself kept those accomplishments and dreams beyond my reach.

This "not-enough-ness" manifested itself in many ways throughout my life. Every achievement was marked by inner fear that if people really knew how insecure I felt, how incapable I felt, how unworthy I felt, they wouldn't like me, they wouldn't follow my leadership, and they for sure wouldn't reward me.

"Not-enough-ness" showed up in my relationships. When we don't feel enough, we attempt to fill that leaking bucket of self-worth with affirming friendships. As a young man, people teased me about my "entourage." I never went anywhere alone. I was always in a crowd, and most often, I was the clown, the popular one, the loud one, the fun one. Like a hamster on a wheel, I was running fast on the inside, working for validation and simultaneously dreadfully afraid that if my friends really knew me, they wouldn't like me. Why? It was obvious to me. I wasn't enough. But, I was determined to "earn" approval through my "people-pleasing" pattern.

This showed up in my dating relationships. I needed lots of affirmation. In Gary Chapman's beautiful work on the "love languages," he identifies one love language as "Words of Affirmation." That was like oxygen to me, or as I will later describe it, like cocaine. If someone praised me or validated something I did or seemed to like me, I was drawn to them.

What I didn't know then—but which I clearly see now—is that I actually chased this affirmation, believing that my worth and value would somehow be imparted to me by others. I thought it was an outside-in experience. Now I understand that it's inside-out.

See, when it's an outside-in experience, the responsibility to keep me pumped up, keep my ego stroked, and keep my feelings of worth buoyed was on someone else. It was "their" responsibility. It wasn't on me. It was on them. And if they didn't fill my self-worth bucket, I'd find someone who did.

How terribly unfair of me. Further, how terribly mistaken I was. And further, the feelings of unworthiness within me also prompted me to seek validation from achievement.

In education, high school graduation wasn't enough. I had to go to college, becoming the first member of both my mother's and father's families to graduate from college. But that wasn't enough. I had to get my master's degree and then my doctorate.

As a runner, I wasn't satisfied to just complete a race. If I wasn't in the top tier for my age group, it was a failure. And 5k races gave way to 10k races, which yielded to half-marathons and then to marathons.

In martial arts, I trained for seven years, aiming to earn a black sash in a very aggressive style of kung fu. Just being proficient in self-defense was not enough. A serious back injury stopped my climb at the brown sash level.

Then came scuba diving. Getting the entry-level certification of Open Water Diver didn't fill my self-worth bucket. I had to get the Advanced Open Water certification, then came Rescue Diver certification, and finally, Divemaster.

What's behind such predictable behavior? A man living his whole life trying to fill a self-worth bucket with external achievement, relationships, and "people-pleasing." And the net result? My self-worth bucket was still empty. However, my misery bucket was filled and splashing out. If only I had paid attention.

SECRETS

In addition to my "not-enough-ness," my childhood taught me to keep secrets. My parents both grew up believing that there is a public life and a private life. My brother and I were told about many facets of our lives, "*This is a family secret, and we don't talk about it with anyone.*"

An example of this was that while driving to church, my parents would often fight. My brother turned inward and was stoic. I cried. Mom or Dad would chastise me and tell me to "wipe my face" so church members wouldn't be able to detect we were anything but a perfect family. We'd get out of the car, smiles plastered across our faces, and walk into church.

That learned behavior came back to bite me later in life.

YOU CAN GET BACK UP

One of the greatest gifts my parents gave me, in addition to the best love they knew how to give, was the belief that *Johnsons get up*. My brother and I both competed in athletics, and we were taught to rally after a loss, get back into the game after a hard hit, or make the next inning, the next quarter, and the next round better after a failed effort.

We lived with a couple of mottos in addition to getting back up. Another was *Play hurt*. Whining wasn't tolerated. Quitting was a mortal sin, and excuses didn't cut it. So, we played hurt. Complaining didn't help, and there was simply no reason to let an injury keep you from competing.

That mindset has served me well, for the most part. The principle of *Bounce* comes from the image of a person knocked down and bouncing back up.

One aspect of it that isn't healthy, however, is always playing hurt. What I now understand is that some hurts—especially the emotional ones—must be acknowledged and dealt with, or the damage inside gets exponentially worse, and the behaviors that come from emotional wounds can be destructive. While playing hurt, I kept trying to quench this thirst for "enough-ness," to fill this inner ache of needing acceptance and hide this pervasive fear of my secret insecurities. Like a duck on a tranquil lake, he seems calm and lovely. Yet, underneath the water line, that same duck is paddling like crazy. I was living this duality.

CHAPTER THREE

A Strength That Left Me Weak

A RISING STAR

I do believe I was created to serve and elevate others to their best life, and I have done that mainly in the context of local churches. These days, my life's work extends beyond serving a local congregation to reaching a broader audience, as a digital content creator and Mindset Coach. With the goal of inspiring, encouraging, and coaching 100,000 people every week to realize their fullest God-given potential, I've expanded my efforts well outside the church.

The difference between my goals now and how I lived earlier in my life is that I now design and define the expectations for my future without consideration of "people-pleasing." I now understand the extreme pressure I allowed and lived with for decades when trying to live up to the expectations others placed on me, while trying to quiet the inner-demons haunting me. This was decidedly a factor in my eventual collapse. And I now live with the certainty that each person is born for a level of greatness and has immense worth and value—including me.

Coming out of my family of origin, and then after a lengthy stint in the educational system, a forty-year career followed. I started serving others in a small church in a rural Kentucky town. I then served another church that grew to a couple thousand people attending each weekend. Then, I served as a

Teaching Team member for a church in California, where I regularly spoke to crowds of 20,000.

Then I made what, in retrospect, was a consequential move, going to serve a church where my mandate was a "turn-around." The church was in steep decline, heavily in debt, and struggling financially. Tactically, I knew how to lead, and I knew how to bring vitality and health to the organization. Within a few short years, the church was thriving. It grew by a couple thousand more people, there was strong financial cash flow, thousands of people's lives were changed in positive ways, and I was winning at the task.

Yet, I found myself mired in an unhappiness I didn't understand. I was the replacement for the founding pastor, who had been forced into retirement by the church Elders because of the decline in attendance and revenue, though he was greatly loved by the congregation.

When I came in, much younger and with new ideas, each suggestion was met with pockets of seething resistance and verbal condemnation. I represented the harm that many believed had been inflicted on the beloved founder. For a person with Imposter Syndrome and with deep-seated feelings of "not-enough-ness," the unrelenting criticism tore holes in my already fragile interior world. I was among the walking wounded, and no one knew it. I kept going, because Johnsons *get up and play hurt.*

Those were the years when I learned a valuable lesson about strengths and weaknesses. I always believed that a strength is something one does well, and a weakness is something that one does poorly. Though these are the classic definitions, they are not always true.

I learned that I was good at managing and leading a large organization: over 70 full- and part-time staff, a six-million-dollar budget, and the deep complexity characteristic of a primarily volunteer organization. Though good at leading within this very corporate structure and team, I was miserable inside. And that's where I learned this more authentic definition of strengths and weaknesses:

A weakness may be something you're good at, but it leaves you empty and unfulfilled. While a strength is also something you're good at, it leaves you satisfied and energized.

I was living in my weakness, not my strength. At day's end, I'd slouch in my car with my stomach knotted and acidic and make the drive home, where a growing urge for wine crawled up my throat.

Far removed from being with people, engaging in their lives, and having time to really listen and walk with them through pain or transition, I was reading spreadsheets and considering debt management techniques while resolving staffing dilemmas. I did the business side well; I just didn't stay well doing it.

If someone landed on my calendar for pastoral care or counsel, they got fifteen minutes. I know they felt short-changed, and I did too. I certainly needed them as much as they needed a pastor.

This only heightened my sense of "not-enough-ness." I recall my internal questions from those days: *Why was I running on empty all the time? Why was my mood circling the drain in deep shades of blue? What was wrong with me? Was I in over my head? Would people find out I wasn't capable and that I was winging it?*

Imposter Syndrome was in full bloom.

It was in that season that I made one of many poor choices. I chose not to talk about what I was feeling. Now I know:

We can't heal what we don't reveal. We are only as sick as our secrets.

Do you see how life in my family of origin gave rise to demons and struggles within me as an adult? I had all the outer appearances of success, but inside, I was a mess. The two forces of my childhood—"not-enough-ness" and secret-keeping—were the chief drivers in my life as an adult.

I fantasized about escape. Psychologically, this is called a fugue state—a fantasy of flight away from the harsh circumstances of one's life. I started to envision a different life, one with no heavy responsibility, a life more suited to my limited beliefs about myself, and a life far from the criticisms that hit fast, frequent, and hard.

As a "people-pleaser," it ate me up to get a critical email or if someone confronted me about how my leadership took the organization in a direction they didn't agree with. I worked hard to explain myself and desperately wanted people to understand. Energy spent "people-pleasing" felt like a little boy stretching his hands out to receive a hug, only to have those hands slapped away.

Then, emotionally, I would whisper: *"Play hurt. Get back in the game."*

A pit we commonly fall into is trying to make people understand. If you are true and pure in motive, you do not need anyone's understanding when you make a decision. I didn't believe that at the time.

Yes, there is value in working together toward understanding. I get that. I'm a believer in consensus building. Yet, behind many of my attempts to get people to understand, I was really doing two things: "people-pleasing" and wanting people to like me. We crave the validation their *understanding* and *approval* provide. This also feeds into "not-enough-ness." We begin to doubt and second-guess ourselves, not really trusting our intuition, our wisdom, or our experience; therefore, we seek outside validation. Do you see the vicious cycle? Trying to please and feeling "not-enough"; trying to please more and feeling even less within ourselves.

Do you recognize any of this in you? If so, hang on! Good news is coming, I promise.

HOW TO CATCH A FALLING STAR

Playing emotionally hurt finally shows up. You can only suppress, avoid, and deny for so long. The hurts from my childhood (including molestation by a church worker that I never spoke about until I was 50 years old), the pressures of leading from my weakness (doing what emptied me), the sense of being an imposter, looking outside of myself for validation, all coagulated and I made decisions that would alter the direction of my life and the lives of those I cared about.

You leave before you leave.

My escape fantasies became more frequent. There's a huge point to make here:

You leave before you leave.

You make subconscious choices before you make choices in the tangible world. What occurs in the mind, what you fantasize about, what you think about, and what you focus on is eventually manifested in behavior. I left my old life, my marital vows, my parenting role, my reputation, and everything I cared about long before I took any outward action. And you will, too. This is why our thoughts become our best friend or our worst enemy.

One day, as I sat at my church office desk, surrounded by a library of my academic books, waiting for my next appointment—in what always felt like an unending stream of appointments—I saw water splash on paper on my desk. Confusion quickly gave way to clarity: I was crying and didn't even realize it.

Depression had fully settled on my spirit and mind. I didn't understand depression at all. I just got up each day and went back into the game. Each day, a piece of me dropped away; emotional parts of me vanished or collapsed. Each day, hidden pain showed up in emotional expressions—usually tears.

Depression, for me, was anger turned inward. I was mad at being afraid all the time. I bristled because outside validation failed to fill me up, I simmered with heat because my responsibilities were incongruent with my strengths, and my face flushed red with anger that the "people-pleaser" in me didn't know how to set and hold boundaries. The resultant depression sent me on a quest to "feel good again."

And because I craved and was addicted to outside validation, I looked outside my marriage for it.

The church where I ministered was located in an area with a strong celebrity culture, where many people sought and craved proximity to celebrities. The church culture was not immune to this—I was not immune to this. On Sundays, after speaking to thousands, I would stand near an exit—sometimes for nearly an hour—as lines of people extended through the cavernous building, each person waiting to shake my hand. It was surreal and heady.

Nearby security guards stood beside me—and as I look back on it, I don't understand why. I was never threatened and never felt unsafe. But I suppose celebrities need security, and I felt a gravitational seduction to that identity. Dozens of times, while standing at the church door, a woman would slip me a business card or a note with her phone number and an invitation to meet privately for coffee or a glass of wine. My moral compass spun, first from shock, then to the inner thought of *that felt good,* then to desire. Finally, the compass spun wildly, losing sight of magnetic north.

A hubris formed inside of me within that culture, a feeling of being immune from consequences, above the rules, and therefore bullet-proof. Within hubris are seeds of destruction, which grow every day until something ugly emerges and consumes you.

The attention and compliments and seeming adoration of others became like cocaine. I got closer and closer to the edge. Small flirtations seemed innocent enough, and I deserved to feel good, didn't I? That was my inaccurate rationale. Then, I crossed more moral lines and was pulled along by the feel-good parts of the brain, the parts that tasted like true validation. The very same parts of the brain that fire from a cocaine hit also fire from promiscuity, and the addiction is just as real. It was thrilling, physically and psychologically. One person later described it for me like this, "Brad, it's like jumping off the Empire State Building. The first 99% is sheer thrill. The end of that, however…"

But—and don't miss this—as I started living below who I was and who I wanted everyone to believe I was, there was a destructive, counterproductive effect. I was getting the physical and emotional high from great attention and what felt like deep affection, yet I was getting hammered in my spirit from guilt and shame. It's not overstating it to say I was at war with myself.

Self-respect is the elixir of self-esteem. You start to lose it when you live below your own personal standard of conduct, fall short of your desired or stated values, and break your promises to yourself. I fell into even lower self-esteem. I found myself wallowing in self-loathing.

Science has shown that our thoughts have energy and frequency. What my thoughts were sending out was erratic, and what my emotions emitted was shame and guilt. I was an erratic, shame- and guilt-riddled man on the brink. Looking in the mirror each day, I barely recognized the face staring back. Dark circles encased my eyes, and a hollowness shrouded my features.

Each day became more difficult to endure. My duplicity and hypocrisy validated everything I had ever thought about myself: *I told you that you aren't worthy, told you that you are an imposter, told you that you don't deserve this good life.* Those thoughts then also fueled my desperation to get out of the pain and cycle of behavior where I felt trapped. It felt like I was holding a ravenous wolf by the ears. I wanted to let go but was afraid of what would happen if I did.

Have you ever felt like your life was circling the drain? There was still the old me in there, but he was dying every day. There still was a man who valued being faithful to God and family and craved to once again be so, but my twisted guts suffocated him slowly. There was a man who wanted to love

God... wanted to be loved by God, but those thoughts became a faint echo with each passing day.

I no longer knew who I was. I no longer trusted myself, my instincts, my faith. I was leading others, yet I was more lost than I had ever been.

That led to the first attempt to end my life. Each of my attempts involved drunkenness, or as I thought of it, "liquid courage." I recall wanting to die in a car accident, but that time, I apparently didn't even make it to my car. I staggered into my bathroom, emptied some pills I had available into my shaking hand, and dry swallowed them. I don't recall getting sleepy. I don't recall the next period of time. I just remember waking up on a sidewalk far from my house with my dog curled next to me in the middle of the night. How I got there, who saw me along the way... I have no idea. It's a wonder I wasn't hauled off and institutionalized. I felt crazy and had no grasp on what to do. All sense of spiritual or emotional bearing, compass heading, or direction eluded me. Shame sat heavy, breaking even my posture, which felt bent and old. A shiver shook me that night on the sidewalk, not because I was cold but because I didn't know how to remain in the hell I had created.

What is important to note about that first secret suicide attempt is the event prompted a half-hearted realization that I should try to talk to someone.

Shame and the mounting fear of being discovered, then disapproved of and ultimately rejected, was more than I could take. But talking to a friend or someone in my organization felt impossible. I was the consummate secret-keeper, and the grip of the secret around my throat caused a constant state of nausea and anxiety.

I sought out a therapist. He was in another city and was nationally renowned. He let me ramble out my story and, after a session or two, put me in touch with the seething anger that he believed fueled the patterns of my thoughts and behaviors.

I caught a glimmer of understanding about one of the drivers, but I didn't change my behavior and continued to spiral downward. I remember having thoughts and even flashes of insight where I was lucid enough to see that there were brief opportunities—like off-ramps on a freeway—where I could have extracted myself from the promiscuity and chemical abuses that now had a firm grip on me. Yet, given a way out, I continued on. Believing I could have stopped and didn't is a part of the cycle of shame that nearly killed me then and that can still occasionally knock on my door, even this many years later.

Much like a dilapidated person, bone-thin and barely alive in a crack house, I was disintegrating at a soul level in the church house. I kept seeking the cocaine-like highs—as temporary as they were—from a selfish set of choices. My energies were trapped in these harmful choices and the emotions attached to them. I tried to numb my feelings through the overuse of alcohol and abuse of prescription drugs.

A second suicide attempt left me in a state of total despair. I awakened in the backseat of my car, far from home, not remembering how I got there... far from anything and everything I had once been. I sobbed for what seemed like hours. Another failed attempt at an overdose. My body spasmed, racked with all the pent-up emotions that drowned me in a tsunami of pain. It wasn't a cleansing cry. The sobs were shovels of dirt heaped upon me, burying me in a self-dug grave.

My thoughts were for my family: I was desperate to hide from them the person I had become and spare them the emotional devastation that I knew would hit if they discovered the state I'd gotten myself into. Where was that thought before I spiraled downward? I know it was there, but I was hell-bent on running from my demons, and in doing so, I became someone—something—I no longer knew.

Creeping through the darkness was the haunting awareness, a visceral whisper that all this was my doing. No excuses. No amount of anger, no residual damage from a traumatic childhood—none of it—could account for my choices and actions. My soul died in the backseat of that car. I realized completely at that moment that every semblance of the life I once had was forever and completely over. At many levels, I had, in fact, already successfully *killed myself.*

I sleepwalked through the next weeks, considering what to do. Each Sunday, I showed up to teach messages of hope and renewal to a crowd of thousands while completely lost in the hypocrisy of my inner darkness.

I had a green room (celebrity perk) where I'd sit on comfortable furniture, with a snack bar fully stocked and a security guard outside my door. In those weeks, I would collapse behind that closed door and weep. Then, just like I did in my parents' car during my childhood, after hearing Mom and Dad fight all the way to church, I would wipe my eyes, paint on a smile, and march out and do my duty. Life's patterns repeat until we address them.

Ironically, church leaders would say during that period that I was delivering some of my very best teachings. I can only assume God was protecting the congregation from the truth of me.

IT WAS TIME TO CONFESS

In therapy, I came to understand that delay in the inevitable would lead to far more destruction than I already created. I couldn't undo the past, but I could, as my therapist suggested, mitigate the level of future compounded harm. I decided to confess to my wife and daughters. I will never forget those moments. It was in their eyes and the cries from their hearts that the full weight of what I had done descended. If it was nearly unbearable for me, it was far worse for them. They each were innocent in this nightmare.

I have blocked out many of the details of the conversations and emotions that followed this disclosure. What I do recall is that I decided I needed to resign. My wife and I didn't know where our marriage would or could go, but we thought if I could resign and geographically move away, then perhaps we could spare the church some of what would happen if or when news of my life broke. That was not to be.

In one sense, my story isn't one of a person who got caught and then feigns remorse. I never got caught—not in the commonly understood sense of that idea. My deep, aching remorse came before exposure. I was, however, fully caught in the throes of damning consequences as I watched my children and wife torn apart by the shrapnel of what I inflicted.

In the spring of 2007, I resigned from my position, citing burnout and depression. That was only a partial truth, which is a full lie. People were confused. I was numb, and my family was broken.

Being away from the church took away the only external restraint that I'd had on my life. Without this holding me back, I dug deeper into destructive choices, and my abuse of alcohol and prescription drugs deepened—to the point of having a couple of people get prescriptions in their names so they could give me the pills, anything to dull the pain and shame. Someone wisely noted that some people change for the better when they hit rock bottom. Others hit rock bottom and then dig a basement. I fell into this latter group.

My therapist, a new one who spoke in a way that made sense to me, said, "Brad, you are driving as fast as you can toward a brick wall." I felt that. I was accelerating in the wrong direction.

Another error, piled on top of a mountain of many, was the anger I pointed at others—my wife, the church, Christians in general, my upbringing, and my abuser (none of whom were responsible for my choices and my behavior). Though I had earlier determined that I was the one to blame, as I fell further away from my spiritual center, I once again tried to assuage my pain by directing anger at others. Please, it's important that you hear this: It was all on me. This was my doing, and any attempts to turn anger and blame on others was only deflection on my part. It tragically delayed my healing.

This disastrous cycle went on and on. Anger directed outward fueled the heat of anger I turned inward. And just like validation can't come from outside us, pointing our anger outside of ourselves is also unhelpful. There was now just more anger to deal with.

And I didn't believe I could turn to God. Spirituality had been a source of strength and help throughout my life, but old tapes from childhood—about God being angry, God catching you when you're "bad," God "getting you" when you disobey, etc.—all combined into a concrete conclusion: I had gone too far and done too much. The God I learned about growing up had certainly written me off, just like I had written me off. God and I agreed about that. I was finished.

The fall of a star is a spectacular scene to behold. What we see when a star is falling is a star literally dying. People can't look away. It's captivating. We often don't associate it with death, but that's exactly what's happening. Many people have said to me through the years, "Brad, we didn't know what to do, but we couldn't look away."

I understand that.

My family disintegrated in the dying embers of the star's fall. To see that happen because of my life was the most excruciating experience I have ever known. My children could barely acknowledge me. I lived ashamed, and I emotionally separated further.

Financially, life became a nightmare, and I was shocked at how fast and hard that tsunami hit. At one time, we were wealthy, in the top one percent of income in the country for my profession, but the money evaporated at a frightening clip. I filed for divorce and sobbed the day I knew papers were served to my wife. My oldest was off to college, and my youngest was desperate to begin her freshman year in college, away from the hell I had created. There was nowhere to turn, nowhere to hide, nowhere for me to escape—only one option left. Suicide was my answer.

A RANCID SMELL IS HARD TO HIDE

A few months after I left the church, and now more lost than ever, a church leader called and said other leaders wanted to meet with me to discuss rumors they were hearing around town. By then, I had become flagrant in my behavior.

I agreed to meet. Details of that meeting are sketchy because I went into it high and foggy, trying to numb with chemicals what I knew would be awful. I have fleeting memories of what they said and the anger they expressed. I remember thinking, *Me too. I hate me, too.*

Whatever I said or confessed in that drug-addled state was soon shared in public meetings at the church. The church I hoped to spare was devastated. It's hard to calculate the pain one person with a string of terrible choices can inflict. It was awful. And in the residue of all that tumult, any fondness people once had for me died. My thirty-year career vanished. A church, a family, and a man were burned to the ground, and for what? Because I wanted relief from the pain and fear I carried for years? I played the fool, and now it was all so terribly sad.

None of this is a play for sympathy. I had once been at the top of the profession. I was booked three years in advance to speak to crowds of thousands and was paid thousands per event to do so. I had signed my first book deal with a generous advance. I was known and respected. I had influence. I was helping people. On a graph, my career had a sharp trajectory of up and to the right. But when I fell, it was shocking how fast it all crashed. The book deal fell apart, the purpose of my life was gone, and everything that once mattered was gone. At the most significant level, my family was gone.

All of this is a predictable conclusion to where my mind, my beliefs, my words, my emotions, and my actions took me. This is crucial. When my mind, beliefs, words, emotions, and actions were aligned toward self-destruction, it happened with horrific swiftness and power.

Please pause and absorb this.

The same alignment pointed in a better direction can have proportionally the same power for good.

Those same areas aligned toward restoration and healing are what helped me *Bounce* back—but I'm getting ahead of myself.

When I found myself unemployed, divorced, estranged from my children, and publicly despised, I researched what combination of drugs was found in the body of actor Heath Ledger after he died. He was appreciably younger than me when he tragically and accidentally overdosed and died. I determined that someone my age would certainly die if I took the same combination of drugs. In my mind, I could see all the hurt I had inflicted on the people I loved. And the glaring contradiction that I loved these people was stark, considering my awful choices. I was beneath rock bottom. I stopped digging deeper. I had nothing left. It was time once and for all to end this nightmare for everyone.

    I attempted for a third and final time to take my life. This time, I was sure I'd succeed. Problem solved. But as it turned out, I even failed at failure.

CHAPTER FOUR

What to Live For

*Waking up from my third suicide attempt was the
most awful, beautiful moment of my life.*

Can something be both excruciating and revealing? Awful and illuminating? Yes!

It was awful because I was still waking up in the world of my failure, still in the world where the shrapnel of my choices left many reeling and spiritually bleeding out. I didn't know how to keep living in that world. Shame, like strong hands, choked me.

It was also the most beautiful moment of my life (and I only see this in hindsight) because that set me on a course to begin believing there was a destiny for my life and that God had a plan that transcended what I knew or believed about myself.

I was once asked when a person can see the furthest. I suggested an answer to that question. I imagined a person standing on the summit of Mount Everest, looking out on the vast Earth beneath. But my friend said,

*You actually see the furthest at night because in
the darkness, you can see the stars.*

Through the patience and wise efforts of my therapist and my brother, I started having conversations that allowed tiny pinpricks of starlight to penetrate what had become a couple of years of incredible darkness. With professional guidance, I began to slowly untangle the mess of my life, the twisted and misguided beliefs that had been the operating system of my life. These efforts revealed my broken beliefs—about myself and my future—which brings up a huge point. It was the belief of others in me, more than my own (which was non-existent at the time), that helped point me toward a future.

When someone believable believes in you,
you begin to believe in yourself.

Sometimes, you must lean into the belief of someone else because you don't have that belief within you. Slowly, their belief begins to seep into your interior world and shows up in tentative believing thoughts in your mind. This is the process that slowly injected the smallest flicker of hope—the first I'd felt in a couple of years—into my spirit. I allowed my mind to believe—in brief momentary snaps—that there could be a future for someone like me. Such a thought had only recently before seemed "impossible."

This taught me the value of having the right people in proximity. *Bounce* doesn't happen in a vacuum. If a ball is deflated, like my mind and heart were, I needed the help of others to inject the air of hope so I could *Bounce*.

I'd like to say life became magically better. In fact, my spiral continued, but in other areas. Financially, I was broken. The million-dollar house my ex-wife and I owned sold at the bottom of the real estate crash in 2007–2008. I had no money and no place to move. I was 17 days from homelessness, and if not for a person stepping forward to help me have a roof over my head, I would have lived on the streets. All I could manage to cobble together was part-time work. Something in my psyche, absolutely at the subconscious level (maybe even soul level), got me out of bed to go job-hunting day after day. Maybe it was the mantra *Johnsons get up.*

A realtor I knew was handling foreclosures for a bank (there were lots of those during the market crash). She needed someone to go to an area of Southern California that no one was willing to go to because it was populated by notorious gangs. Many homes there were in foreclosure, and the bank needed pictures of the properties. I was paid seven dollars per home to knock on the door of gang members and ask to photograph the house the bank was taking away. I did it gladly.

An upside of wanting to die is that you lose your fear of death. I was the perfect person for the task. And in full disclosure, I didn't have one negative encounter. I attempted to be compassionate and kind. And each person, though a bit skeptical, allowed me to do my job. I truly could relate to their predicament; I knew what it was like to lose everything.

The money wasn't enough to live on, and I found myself racking up more debt on my credit card and falling deeper into the belief that this was now my life and it wouldn't get better.

I'd get home at the end of the day, wind-whipped from riding my motorcycle (I couldn't afford gas for a car) house-to-house taking repo pictures. One evening—I'll never forget it—I was very hungry. I knew I hadn't been to the grocery store in a while because of lack of money, but I hoped I could forage in the cabinets for something. I shuffled into the kitchen and scavenged through each shelf and through the refrigerator. All I had in the whole house that was edible was half a bag of Cheetos and about two shots of Tequila. Oh well... dinner.

Sunken into the worn chair in my den, I heard the mail truck stop and then leave. For everyone who has ever dreaded getting yet another bill you can't pay, you know the feeling I had each time the mail came. I often went days before collecting it because I knew I couldn't pay the bills anyway. Something tugged at me that evening to get up from my shadowed mood and gather in the mail. I pulled myself vertical—more emotionally than physically—and brought in my mail. With fingertips stained orange from the Cheetos, I rifled through the bills.

A letter dropped out from all the way across the country from a friend on the East Coast who I hadn't heard from since *my fall*. Front-page scandals have a way of spreading rapidly. Inside was a note that simply read: "Figured you could use this." There was a $75 gift card for a grocery store.

Now, remember, I had just polished off the final bit of everything I had to eat and drink. Like Old Mother Hubbard, in the classic children's story, my cupboards were bare. This gift came right on time. I cried at that person's kindness and cried at the realization that someone like me was still finding grace in the mess of it all. This is a principle I have learned repeatedly.

> ***Everything I need is available and coming my way.***
> ***And it comes just on time. Not always my time, but on time.***

Believing that and seeing it work over and over again bolstered my faith and belief in divine provision.

That wasn't the only time I was hungry. Food scarcity was my new daily reality. I still had the nice clothes I owned and a Costco membership with a few months remaining before it expired. If you are familiar with Costco, you know that during the day, they offer small samples of food—the latest cracker, slice of cheese, or other foodstuff. I would dress as nicely as I could to hide my stark condition and stroll the aisles, take a sample, and quickly eat it. Then, head hung in shame, I would circle the cart, pass by again under the stare of the clerk, and take another sample. I took the hit of embarrassment to address the gnawing clench of hunger. Many days, Costco samples were my only food.

There was other provision. Piggybacking on the principle that the belief in believable people is important, I must mention Russ and Jacquelyn. At a low, low time, I received an email from someone who I didn't know, but who said he had attended my former church and his family had been positively impacted by my influence while I served there. Then he wrote, "I thought you might need a friend." He included an offer to meet. I needed a friend in the worst way.

Though I could write a whole book on the friendship that has developed between me, Russ, and his lovely wife, Jacquelyn, it's important to note that they walked into my life when so many others walked out. And more, they believed I could *bounce* back. Their belief still encourages my heart. Russ shows up often in the rest of this book.

FROM SERVING COMMUNION TO SERVING COFFEE

Another part-time job I secured was at Starbucks. I was attracted to a position with that company because, at the time, even part-time employees had a path to health insurance. Though I had moved to another city to try to reduce the impact of my actions on those hurt by them, the only Starbucks location with an opening was right back in the middle of the area where my former church was. I used to write sermons in the Starbucks where I was eventually hired. And for a year,

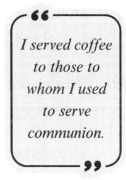

I served coffee to those to whom I used to serve communion.

> *I served coffee to those to whom I used to serve communion.*

I trembled with fear, and my stomach roiled with nausea, considering what might happen as I worked in the epicenter of where I had wrought destruction, but I needed the job. In the Starbucks system back in those days, the first task a new employee was trained on was the cash register. It's the most customer-facing task in the store. And you come face-to-face with every customer.

It was like an emotional Russian Roulette. During my year there, some customers walked in, recognized me, turned on their heels, and walked out. My self-worth reeked of the trash I still believed I was each time that happened. It validated that I was and always would be a loser for whatever time I had left on Earth.

Others used the opportunity to unload their pain, disappointment, and anger. I understood all those emotions. I still feel the burn on my face as other employees and customers would try not to watch, as person after person would tell me what a worthless piece of crap I was. I didn't disagree, but the blunt hammer of their words nearly broke me over and over again. I was sure I'd be fired as this scene replayed many times. And in my confused spiritual state, I fought to reconcile what I believed I deserved from them and their claim to be representatives of Jesus.

Also, that year, a precious few offered me kindness and love. It didn't take many of those to help my wobbling spirit steady and my fragile soul stay intact. Tears burned behind my eyes each time grace was extended. Those days were a whiplash of emotion.

In the meantime, in 2008, an informal team formed around me to help guide me back to health. Soon, the team was ten people—my therapist, several pastors, my dad, my mentor Russ, and others—who agreed to step into the mess of my life. Regardless of the scandal, regardless of what others said about them, these brave, gracious souls began a restoration process. Like a piece of broken, discarded furniture in the hands of craftsmen, this team saw potential for me that I could not see. I didn't meet with all of them all at once. Some were across the country, and we would meet by phone. One local pastor was especially present throughout that year. Other meetings would occur at coffee shops or over lunch—their treat.

I was in contact with parts of this team every week—and many weeks it was every day—with one or more of them for a full year. Additionally, for two years, I saw my therapist individually every week. I just kept loading my credit card, and he graciously gave me a discount.

I had no confidence and didn't trust my own judgment or ability to make a good decision (because I had made so many terrible ones), so I leaned on this team to guide my life. This team asked me the hard questions about what

I wanted in life and, most importantly, *who* I wanted to be. They checked my motivations, challenged my abuses of prescription drugs and alcohol, they asked about friends and the use of my time. They probed how I was thinking and inquired about my choices.

The fatigue I carried laid over me like a winter coat in summer, and it was stifling. Yet, I knew, somehow and for unclear reasons, I needed to progress. I desired to once again be a man my children could love and respect. I repeatedly begged God for this. I did all I could do, which mainly meant I kept my appointments and blindly followed their lead. I had little else within me.

Over that year, that team's unwavering belief that I had a future and that it could be good grew inside me. I literally felt and watched my thoughts turn from hopelessness to disjointed pieces of hope. I put together a good day, then another. Of course, there were hard days and days I didn't fare as well in my thoughts or my actions, but the overall trajectory of my life started improving.

Simultaneously, I worked at consistently being the man I truly knew I could be again, but whom I had lost along the way. I worked at this in front of my children and tried to even show some of this to my ex-wife, who was doing extraordinary work herself and for her future. It was her work on forgiving me that changed not only her life but the lives of my children and my life, too.

Her forgiveness helped her on her healing journey. As my children watched their mom heal, I observed that it helped them while they worked toward their own healing, and it ultimately helped them open up to healing in their relationship with me.

Forgiveness in the spiritual, quantum realm creates the perfect energetic environment for the physics of Bounce to exist.

You never forget the people who walk in when everyone else walks out. I will never forget the kindness given and the risks taken by those who walked in and stayed in my life and walked me from my burial mud to the light. My gratitude is unending for each of those individuals.

Early in this restoration process, I was asked what my goal was at the end of it all. I knew deeply what I hoped for: I wanted to be intact. I wanted my emotions, spirit and mind to be aligned and moving in a direction congruent with who I truly was. And as I said above, I wanted to be back in the lives of my children in a healed and whole way.

What I didn't want was to re-enter formal church leadership. I became so unhealthy under the pressures of that, but more, I knew I was disqualified from that. I'd been around churches enough to know *you don't bounce back from what I'd done.*

I didn't see such a future, and I didn't believe it was possible. We each can create our own reality here. If you can't see it or believe it, it will not be. And I was good with that.

If I could once again be the honorable man that I knew I had been and desired to be again, then I would find some form of work and settle into serving people from the sidelines, anonymously, and out of the spotlight.

The members of my team had other plans. Late into the process, in 2009, two years after my fall and after a year of intensive work, scrutiny, encouragement, and challenge, a couple of members of the team around me said, "Brad, your gifts and abilities are still within you."

That was a shocking concept to me. I believed I'd forfeited all of that, broken all that in irreparable ways. Yet, here they were, right in front of me, insisting that everything I once was—good, productive, and influential—was still accessible and more. It was all right inside me. It blew my mind.

It was mystifying to consider that everything I needed to bounce back was already tucked and packed away inside me. They taught me that my value and my future were already supplied through what was in me. As one who always went looking to "find" what I needed to succeed or be happy or to have a purpose, I began to understand that it wasn't going to be found outside of me; it just needed to be accessed.

I asked my team, "So, what do I do about it?"

What did they expect me to do with the gifts and abilities they were so sure I still possessed? They shocked me further when they said, "Consider stepping back into service to others through a church position."

It was the last thing in the world I wanted. I saw my future life as walking on the sidelines, head down, passing the years. They challenged me to believe that I had a greater responsibility with my one and only life. What they said was surreal; it was as if they were speaking about some other person.

> ***Give yourself permission to believe that greatness lies within you.***

Giving yourself permission to believe that greatness lies within you, believing that there is a future for you no matter the past, is a very hard sell. Yet, when you cross that mental barrier and allow yourself to hope and then believe it's true, *everything* starts to shift in a favorable direction.

Many conversations were spent on this topic, batting around my objections and their belief in me and in my future. Truthfully, terror seized me, and I was racked by panic attacks. I had night tremors and night sweats, considering their ideas. Yet, something tugged at me to stay at the table of this troubling conversation.

Finally, after a month or two of hard truth-seeking, we all agreed that I would be "open" to serving in a church, provided I was not the main leader and the position was outside of California. Both conditions were to protect my mental health and were more palatable to me than becoming a key leader back in California. This was likely the only type of position someone with my record of failure could hope to secure.

With this agreement in place, I dusted off my resume and wrote a detailed cover letter about all the facts of my fall and subsequent journey. Members of the team helped by sending my information to their contacts in churches across the country. I jokingly said (but only kind of joking) that the rejection emails came back faster than these churches even had time to read my information. And each consistently replied: No. No. No.

This validated what I believed to be true. There wasn't a future for me in church work. Though the team could see a future for me, I didn't believe it. This truth of my own beliefs standing in my way, blocking my destiny, was only a kernel of understanding at that point.

Outside pressures mounted. I was still drowning financially. I couldn't find enough substantial work. On more than one occasion, someone from my former congregation torpedoed opportunities I pursued. I felt old patterns of anger rising. It was in the crucible of encountering unkind people who had genuinely been hurt by me and now felt they needed to inflict pain back on me that I learned one of life's great lessons:

I must be merciful to those who are unmerciful.

The response I give to people, *all people*, is completely my responsibility. This truth was born in a crucible of struggle. It was not voluntarily concluded on my part but rather divinely wrenched out of me. But once I stopped living like a victim who had to flail back at an attacker, my spirit calmed, my vibrational frequency rose, my mind settled, and I had access to more energy to pursue a better way.

After months of sending resumes (lots of resumes), a gracious church in Denver, Colorado, interviewed me multiple times by phone and agreed to bring me on staff in an associate position at one of their satellite campuses. Perfect. I wouldn't be a central or focal leader; I could serve people, and it was out of California. It was all coming together—or so it seemed.

Then, the most remarkable thing happened.

That summer 2009, I was working an afternoon shift at Starbucks when my 23-year-old boss said, "Brad, someone spilled a Frappuccino in the lobby. Please mop it up."

"Of course," I said. It was sometimes hard to imagine that this was now my life. If I wasn't mopping the lobby, I was cleaning the toilets.

I grabbed the mop and headed to the lobby. I felt exposed in the lobby, vulnerable to the next angry person who might walk in. While there, I noticed the trashcan was full, so I pulled out the bag and replaced it with a new one. I mopped the floor, wiped a couple of tables, and restocked the sugar packets and stir sticks.

Just then, from the corner of my eye, I saw Bruce Jenner walk in and head to the counter. It was such a rush. Bruce was one of my sports heroes. I watched every event he competed in during the decathlon at the 1976 Olympics. He was the greatest male athlete in the world, bringing home the gold medal. I had once eaten Wheaties cereal from a box with his picture on the front. And here he was at my Starbucks. It was the highlight of my day. But remembering why I was there, I quickly busied myself with my tasks.

A few moments later, I saw Bruce leave. Then, within seconds, someone tapped me on the shoulder. I flinched. As described earlier, I never knew what was coming when someone came in and tapped me on the shoulder. But this time, it was Bruce Jenner. I was really startled. He had walked back in.

"Pastor Brad?" he asked.

"Well, I'm not a pastor, but I am Brad."

"I've been looking for you for six months," he said.

Dumbstruck, I stood and simply listened. Realizing I had lost my breath, I inhaled sharply.

He continued, "My family and I used to attend the church where you served. We never met you. We never wanted to stand in the long lines of people waiting to shake your hand each Sunday. And no one knew where you went after you resigned."

I stood with my mouth gaped open as I tried to absorb all he was saying. Logic and rational thought escaped me.

"My wife Kris tried to find you, and then she asked me to find you. We've been at it for a while now. I finally heard you were at a Starbucks, and I've been to every location in this region. I found out you were here, but I kept missing you. I almost missed you today because I wasn't sure it was you."

"It's me," I said, feeling the need to say something.

"Well, here." He handed me Kris' business card. "My wife wants you to call her. She has an offer for you. Do you know who my wife is?"

"I'm sorry, I don't," I said with some embarrassment. In those days, I never watched television and wasn't aware of the growing fame of Kris and her family.

I looked at the business card. I didn't know what to say or what to do as I juggled the trash bag, wet towel, and mop that I was holding. Finally, I muttered, "Thank you!"

He nodded his head and said, "Give her a call. She'll be glad I found you." He turned and walked out the door.

I tucked the card into a pocket and wondered what this all meant. I had a tingling spiritual sense that something more than I expected awaited me, but I couldn't define it at that time. What I did experience was this truth:

You are one relationship away from your whole life changing in a positive direction.

Before I go further, let me ask you to do an imagination exercise with me: Imagine yourself in a Starbucks apron (green) in the lobby of a Starbucks location. Imagine yourself mopping and replacing sugar packets. Now—and here's where life gets dimensional instead of linear—simultaneously in your imagination of this scene, pull back from it so you are outside of yourself, looking down on what's unfolding. From your elevated perspective, you can see a car pull into the parking lot. The person in the car has been driving at least 40 minutes to get to that parking lot. Bruce Jenner exits the car and starts walking toward the Starbucks. From your vantage point, you see him coming, and you are also able to see yourself inside, busily cleaning up, completely unaware that A.) Help is coming, and B.) It's coming to you.

Here's the point of the exercise: At any given moment in your life, what you need—what you need now and what you need next—is coming to you. You're busy and unaware, but from a place deep within, imagine living your life filled with expectation and faith that what you need is *already* on its way to you. Practice this exercise again. And then, let's resume the story.

That evening, I went back to my apartment where my youngest daughter was staying for part of her summer. We were slowly reconciling, and I was so grateful. When she asked about my day, I was shocked to learn she knew who Bruce Jenner was. She explained to me everything about their television show, *Keeping Up with the Kardashians*. That left me with more questions than answers.

I decided to call my mentor, Russ. I explained what happened at work and the conversation with Bruce and then queried his opinion about what he thought this meant. He said his wife had watched some episodes of their television show, and he believed they owned a deli. They didn't own a deli, but a deli was prominent on the show because it was next door to a clothing store the family owned. But at that moment, he and I both concluded Kris was going to offer me a job at her deli.

"If she offers me ten dollars an hour, I'll take it," I said, since I was making $8.25 an hour at that time, plus tips.

I called Kris. She was amazingly kind and gracious. She invited me to come to her home the next Monday.

I left my shift that Monday afternoon, a stomach full of butterflies, and made my way to the exclusive enclave of homes where Kris and her family lived. My white shirt had brown splatters from hours of serving coffee. I reeked of coffee beans and fatigue. My feet ached from the long hours I'd been on my feet during my shift. Feelings of "not-enough-ness" swallowed me. I felt like I should go around to the back of the massive house and enter through a less conspicuous door.

I was greeted at the front door by one of her team members and led deep into the expansive home to a family room. Shortly after, Kris came in, greeted me with a warm, contagious smile, and offered me a seat. It was the first time we ever met. She possesses what many call "star quality." She exudes confidence and a posture of assurance. She was a far cry from anything I felt about myself at that moment. I felt ugly, undone, not put together, and thoroughly like a fish out of water. I couldn't imagine that I had ever been her pastor, and I wondered if I had made the right decision to come.

She explained much of what Bruce shared but added her personal experience of believing this is what God asked her to do. She said, "I know God spoke to me and told me to find you and to make you this offer."

This is a big build-up for a deli job, I thought.

Kris continued. "We need a church in our area that believes in second chances and fresh starts. So many churches say this, but everyone in those churches pretends to be perfect because of the judgment they know will be heaped on those who mess up.

"Who better to lead a church of second chances, where people can find a safe place to explore faith, and they don't have to pretend to be perfect? You've lived through it." She went on, "I don't know if you're supposed to accept this offer—that's between you and God—but if you do, I feel I am supposed to help you get started."

I was so taken by her sincerity, and by the surprise of it all, the waterworks began. Tears spilled freely. I was so touched by her graciousness. After so many rejections and having for so long believed that most would never give me a second chance, this particular act of kindness was profound.

It also left me deeply confused. "I don't know what to say," I began, "except this Thursday, I fly to Denver to accept a position at a church there. I can't imagine why the timing of all this is happening now, except perhaps to encourage me. And believe me, this is so encouraging."

Kris was completely gracious and understanding and said, "Please let me know how things go in Denver. I'm very glad I found you."

We both had a glint of tears in our eyes. It was a tender moment. I thanked her profusely, got in my car (hoping it hadn't leaked oil on her driveway), and drove back home. My mind and heart were spinning. I called Russ from the car. He was also struck by the timing of it all, but with no further change in our thinking, I left on a flight Thursday morning to the new dream that I was sure was coming true in Denver.

DEATH OF A DREAM

Within the principle of *Bounce,* there is this reality:

Balls don't always bounce in predictable ways.

This is true in the life of faith or the Law of Attraction, or by whatever name you know this process. We *can* bounce back, we *can* bounce up, and we *can* bounce better, but sometimes in a direction we aren't expecting. This is why, when we conceive of a future or when God plants a new dream in our heart, it's important to hold fast to the big picture of the dream but allow flexibility with respect to the details of how it's going to happen and with whom.

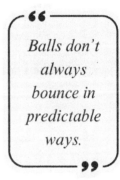

> *Balls don't always bounce in predictable ways.*

I assumed Denver was my "how," and the gracious people I encountered in this terrific church were my "who." I was wrong on both counts.

When the team from the Denver church picked me up from the airport, it was immediately clear from their vibrational frequency that something was wrong—it was that feeling you get when something is very right, or something is very wrong. I had that sinking feeling that something was terribly wrong.

I finally inquired what was happening and they laid it all out for me. They were glad I was there. They needed me to come on board, they confided, because the church had suddenly found itself in a mess and in want of more leaders. My heart sank to my stomach. I felt as though I might vomit. Sweat ran in rivulets down my spine.

They explained that the founding head pastor of the church had been terminated the previous Sunday, and he was taking about half the congregation and starting a new church not far from this church's location. Further, everyone was in an uproar, and battle lines had been drawn.

Immediately, I knew this was not the move for me. At that moment, I didn't believe I could even stand up if I needed to. My legs felt like melting butter. And I for sure knew I could not enter an emotionally volatile season in a church. Coming out of the most broken place I had ever been and having traveled through the most excruciating journey of my life, I knew my next step should not be right back into upheaval. There was no second-guessing on my part.

> **When you know your emotional world intimately, decisions like this become easier.**

However, the disappointment of it all was very real. That was like the death of yet another dream. I had formed an attachment to the plan and the people. Rather than just keeping my eye on the big dream—my *Bounce* back—I allowed my eye to drop down to the details. Again, when it comes to imagining a future and then taking steps to create that future—to manifest it from imagination into reality—it's vital to keep the focus on the end game, the big dream, not the details.

From a hotel room near the airport, I called Russ and shared all that had transpired. After a moment of silent consideration, he asked, "What did Kris mean she'd help you if you decided that starting a church is what you were supposed to do?"

"I don't know," I answered honestly. "I wasn't going to do it because I was going to Denver. I didn't ask her."

"Why don't you call her, see if she'll meet with you again, and ask her," he counseled.

I did so. Kris and I set a time to meet the next afternoon, one week after our previous meeting. That was mid-July 2009.

Many thoughts, reservations, and fears swarmed my spirit as I flew back to Southern California. What was right? What was best? What was meant for me, and what was not?

What I did know is that in my prayer time, a story from the Christian scriptures came to me: an episode in which a woman poured a very expensive bottle of perfume on the feet of Jesus in reverence for him. We are told in the ancient text that the worth of the perfume was one year's salary. This thought was a premonition, though I didn't know it at the time.

The next day, I drove to Kris' home, and again, we sat in her finely appointed family room. I shared what transpired in Denver and then asked her what she meant by her kind offer of help.

Kris said, "I believe I am supposed to offer you one year's salary." It hit me like a lightning bolt. That was the exact price of the perfume the woman in scripture poured as an offering for Jesus. Here was Kris offering such a generous gift to the work of God. Tears once again rolled down my face.

Simultaneously, a conflict burned in my heart. I wasn't going to stay in California. I wasn't going to be a lead pastor. Yet, this offer had God's fingerprints all over it. I recall thanking Kris for her kindness and willing generosity, but knew I needed time to come to peace about my next steps, one way or another.

As I drove away that afternoon:

I became keenly aware of just how wonderfully orchestrated life is when we are honest seekers of direction and truth.

Here I was, finally and reluctantly, in a position to consider a step back into the church world, and then came this resource to help me that had been put on Kris' heart six months before she even found me. And then, I considered the timing of Bruce finding me, the first meeting with Kris, and the death of the Denver dream; it was all remarkable.

Think about it like this:

The resources for my next season of life were already real, present, and available. Further, they were already spiritually headed in my direction, and I had no idea.

And without being open to the possibilities of divine provision, I would have missed it.

We don't know what provision, person, or thing that we need is already heading toward us. We must believe it is *already ours*. Then, it just needs to be accessed. This wonderful truth has played out in my life many times since.

When we take a step, and we don't know the "how," we must fervently believe that all that's required for us to receive a divine "yes" is us giving our yes. And when those two align, miracles happen.

I polled my team of ten from my restoration process to get their wisdom and counsel. Nine of the ten immediately and with great enthusiasm said, "This is exactly what you should do." One team member believed we should stay with the original plan and that it would be much better for me to leave California. He was disappointed that I was considering not sticking to the original plan. I understood his disappointment and felt my own disappointment that he didn't agree.

The "people-pleasing" side of me wanted a unanimous vote, but I had lessons yet to learn. This next one was deeply implanted within me:

You must seek wisdom from others, search your own heart to assess the purity of your motives and thinking, and then come to peace with your decisions.

It is not necessary to have everyone in our circle understand or agree in order to do the next right thing with integrity and for the right reasons.

I took another two weeks to ponder, pray, look deep within, and finally came to a place where I surrendered to the path that had been placed right in front of me. I said yes to a step back into leadership in a church. Though confident this was the right next step for me, my stomach quaked with equal parts anxiety and anticipation. Every entrepreneur and new business owner starting out with only a dream knows exactly what I mean.

It had been over two years since I'd offered my resignation from the previous church. After all that time, I experienced in a profound way the first bounce: *Bouncing Back.*

The journey wasn't easy, and it wasn't fast, but it was worth it. I had yet to learn how to *Bounce Better.* It's important to note that at this early point, not everyone was happy to see me *Bounce Back.*

CHAPTER FIVE

Belief in the Middle of Unbelief

The call came late at night when my phone is usually set to silent. That night it wasn't, and I absently grabbed at it and answered.

"Have you seen the website about you?" a disembodied voice said on the other end.

I rubbed fatigue from my eyes and felt the heavy folds on my brow sink. "What website? I haven't seen anything." A coiling sickness rolled circles in my gut. Life had been tough for a long time at that point. I didn't need it to get harder. My old mindset of always assuming the worst controlled my thinking on that call. Pushing my fingers through bleached white, spiky hair—a nod to my rock and roll, drug-addled former persona—I waited for my friend's information.

"It's..." I felt his hesitation more than I heard it. "... hard to look at." He gave me the web address.

I walked down the hall to the living room, where I last placed my laptop, grabbed glasses off the coffee table, and fired up the screen. I typed in the web address, and there it was. A picture of a bulls-eye target with my face at the center and blood-red letters offered the chilling message: "Kill Brad Johnson!"

I wasn't afraid. In a strange way, I was drawn into the image. It didn't seem strange to me. I had once fully agreed with the premise. I had also wanted Brad Johnson dead. That said, however, by this time, I no longer wanted to die and certainly didn't appreciate someone advocating for my murder.

A friend was also greatly disturbed by this threat and went to work to find the creator of the website. The person who designed the site was technically savvy and hid his tracks very well. It took a forensic tech investigator—that a friend graciously paid for—to track this person's identity. The process took months.

Once he was found, I reached out to him and invited him to coffee. I must imagine he was very surprised that I found him and likely a little unsettled by my invitation to meet. I didn't feel all that comfortable, either.

We met at the same Starbucks where I worked for a year and where Bruce Jenner found me. The friend who helped me track this man didn't want me to meet him alone, so he arrived earlier and took a seat at a corner table to keep an eye on things.

When the web designer arrived, we both ordered coffee and took seats at a two-top in the back of the room along a wall near the bathrooms I once cleaned. I thanked him for coming and then made this offer: If he listened to my journey—of all I had done to *Bounce Back* and how it came to be that I was going to be involved in a new church—and at the end of it he still wanted me dead, then I would leave him be, and he could just run with his website any way he chose. However, if, after hearing me out, he decided to take the website down, I would be most appreciative.

I shared my story. He acknowledged that he didn't know any of it and had put up the website out of his anger for the harm I had done to the cause of faith and the anger of others he knew. At the end of our coffee, he agreed to take the website down. And then, for a couple of years, he was in regular attendance at the new church.

It's amazing what can be accomplished when we take time to talk openly and honestly and take the opportunity to get to know another person. I ended up liking him, and I believe he liked the new me.

THE RESOURCE AND TEAM DILEMMA

Ask any entrepreneur starting a new venture what their two greatest challenges are, and they will say money and people. Having the start-up funds and having the right team in place is crucial to success.

These were the early challenges in my *Bounce*. In the same way that a wet blanket laid over a helium balloon will hold that balloon on the floor, a lack of financial or human resources will put a wet blanket on your dreams. But what we have already discovered is that all the resources you need are not only available, but already in motion in your direction when you believe. You just must learn to access them. I was in a good spiritual place at that point and didn't struggle under the weight of those details.

Here's how it works. I took an inventory of what I already had in place. Most importantly, I had me. I was now fully on board, fully engaged, with a mindset for resilience, power, and abundance. This was a far cry from the not-too-long-before version of me, who was broke, broken, suicidal, and despondent.

> *Emotions are stronger than logic.*

You attract who you are. I was a believer in the new dream. I could see it. I could *feel* the emotions of it happening. And this is a huge scientific and, I believe, spiritual point:

Emotions are stronger than logic.

If our belief about anything (God, our future, the dream) is only in our logic, we have already laid a wet blanket on the balloon of belief that tries to rise. If, however, we engage with our mind and our emotions—we learn to feel something while at the same time believing it's real even before it is—then we are on the path to creating it.

I began to share the vision for a new church with a few others. My relationships were fragile, and my network had shrunk considerably. My closest friends attended a small Bible study that gathered in my rented townhouse (that Russ and Jacquelyn helped provide) each Tuesday. We called it Taco Tuesday, and everyone would bring the ingredients, and we'd combine them, potluck-style. This group could feel my higher, enthused vibrational frequency, and theirs added to mine created an even more energetic dynamic.

Is that an odd concept for you?

You believe this whether you realize what it is or not. You've walked into a room and immediately felt: *Uh oh, something's wrong.* Or you "feel" the tension emanating from another person and wonder what the elephant is in that room. That's because we have electricity in our bodies, and the frequency of the electricity rises or falls depending on mood and other factors. We have felt this from others, and they have felt it from us. It's basic science. Studies conclusively show that the heart emits an electrical frequency that can be measured even outside your body. The same is true with your brain.

It's much like cell phone frequencies. There are an unlimited number of frequencies available. Your phone has one, as does mine. When those frequencies connect, a conversation is possible without wires, without physical mechanics; it all happens as frequencies travel in space between two phones.

What's interesting about this concept is that we are attracted to others who share a similar frequency. Let me say it like this: Positive people attract positive people, while negative people attract negative people. There is something that is transmitted in the space between people that causes a connection, or, in some cases, a disconnect.

My small Taco Tuesday group was drawn to the vision of the church because they could feel my new energy, my new excitement, and the firm conviction this was the next leg of my journey. And I invited them into the dream.

We were small, but we were mighty. There, in my compressed living space, with this tiny volunteer team, the dream for the church took shape and form. Plans were made, details decided, hopes shared. Next, we needed a place to meet.

As promised, Kris Jenner covered my modest beginning salary, and I will forever be grateful. I went from trying to exit this life to being filled with a sense of awe that my life was *bouncing back.*

Without Kris' contribution toward my salary and without the contribution of Russ and Jacquelyn toward my housing rent, the stress of working the four other part-time jobs I once carried to make ends meet would have bled off the energy needed for this start-up. I had a year covered, and I was determined to make the most of it.

Russ and I began the hunt for a location for the church to meet. We looked at dozens of places without success. Generally, the issue was cost. Who hasn't had a wet blanket thrown on top of their dream balloon because of cost?

One day, and I think it was Russ who first saw it, we read an article about new churches starting in movie theaters. Generally, those large, comfortable buildings sit empty and unused until midday on Sundays. After some research, we approached a theater chain in our target area of California. And frequency attracted frequency: They were as excited to have us—and our revenue stream—as we were to have them and their space. The price was a stretch but within reach.

We also tackled the foundational work of determining what we'd do for child care, for music, a sound system, and advertising. The genius of foundational work is discovered in the details. Once again, we attracted like-hearted people who vibed at the same frequency: They got excited at the same level over the same things as us. All our volunteers were lined up, primed, and ready to launch.

When I gave Kris an update on our progress, she inquired what the theater was charging for space. The cost was $12,000 a year. Kris said, "I'll cover the first year." Wow! Again:

The resources for everything you need and all you dream of are already present and moving in your direction when you are operating in belief.

You must step into the future with a confident belief in divine provision and be open to it coming in surprising ways.

With a volunteer team set, the foundational decisions made, a building to use, and a burning vision for the future, we set a date for our first service.

CHAPTER SIX

Love is in the Air

Simultaneous to these other parts of my life assembling in nearly unbelievable fashion, a friend reconnected me to someone I had met nearly a decade before but with whom I had lost touch over those many years. At the time we first knew each other, we were both married and barely had any interaction. We worked for the same organization that had over 200 employees. It wasn't easy to know others outside your team, but we had a vague knowledge of who the other was. Her name was Karen McCann.

Through those years when we hadn't stayed in touch, Karen became widowed, having lost her husband suddenly and tragically. And I, of course, had lost my way. But now, in the process of *bouncing back*, a mutual friend, Gerald Sharon, reintroduced us. The connection was immediate and significant.

We lived two hours apart, so we spent countless hours talking by phone and really getting to know each other on a deeper level. Once a week, I'd make the drive to see her, and on Sundays, Karen would drive to serve as a volunteer at the church. We would have lunch and she would drive back home. This went on for some time, even as the church launched and was up and running.

A deep-seated and lingering insecurity arose in those months. Having once failed miserably at the end of my previous marriage, I was filled with doubt about my ability to have a healthy and lasting relationship. "Not-enough-ness" rose and fell like the tides of the sea. Karen was patient and even initiated at times when she and I both knew I was moving too slowly. I'm so grateful she did, and I was reminded that broken places in me still lingered.

As our love grew and deepened, it was obvious to us—and to everyone who knew us—that such a love was the stuff dreams are made of. Karen made the decision to move to a condo in my area, and I introduced her to Kris Jenner. On a walk one morning, Kris indicated she needed a personal assistant. Karen wasn't looking for anything full-time or time-intensive, but as things went, Karen started working for Kris slowly and soon was Kris' full-time personal assistant for several years. Those two wonderful, strong women have much in common and remain dear friends to this day.

As a matter of fact, Kris got her ordination papers and, along with my dad (who you'll recall was a pastor), helped officiate our wedding at her home. It was filmed and included in an episode of *Keeping Up with the Kardashians*. LeAnn Rimes, who attended our church at the time, graciously sang at our wedding. It was truly magical. Who could have imagined? Life can certainly surprise you in wonderful ways.

Karen's contribution to the church, to my life, and to the future we are co-creating is nothing short of miraculous.

Critics' voices rise in direct correlation to the height of your Bounce.

Remember this principle: *New Levels = New Devils.* A disappointing aspect of the Christian faith is that it is 100% comprised of flawed people. Your organization is too. I am flawed, and so are you. This is why the motto for our church is: "No perfect people allowed." With flawed people come flawed attitudes, flawed beliefs, and flawed interactions.

Many in the church world sincerely believe that people cannot *Bounce* back, or if they can, not nearly as high as they may once have been before a failure or setback, and certainly not better. To say I had detractors is an understatement: Those who didn't believe in me or didn't believe in the vision for a unique type of church that was all about second (and third and fourth) chances. Their open criticism of me and the church was loud and frequent, often convincing those who considered coming to our church not to do so.

And this is the importance of your resilience, empowerment, and abundance in what you think, what you believe, what beliefs you use to create an emotional set-point, what words you say about yourself, your future, your dream, and what actions you take. When outside criticism hits, the alignment of every part of you must be strong and focused.

Remember what we learned earlier:

The answer isn't coming from outside of you. It's already in you. Everything you need is in place in you. This is divine design. God created you in (with) His image.

And because I was no longer living from an outside-in place (seeking or needing validation about me or my future from the outside), criticism no longer had the power it once had in my life. My renewed confidence and belief structure allowed my *Bounce* to rise through the storm of negative opinions and loud voices.

A FRONT-ROW SEAT AS IMAGINATION TAKES SHAPE AND FORM

Let's go on a journey into the mind. When you were a child, you lived primarily from a place of imagination. Trees became fortresses, blankets became tents, cloud formations were UFOs, rocks transformed into horses, and closets were filled with all kinds of dark, lurking creatures each bedtime.

What happens over time is that imagination gives way to logic and memory. Rather than envisioning what "could be," we settle into "what is"—or worse, "what was."

When we live mainly with logic and memory, we keep repeating what used to serve us, whether it still does or not. Pain and trauma from our past are emotional and belief setpoints, and though events may have transpired some time ago, the emotions of those events can remain. It's where many get stuck. They still feel and experience the past.

Think about the impact of that. Emotions lived *now* are based on old experiences, which means we are living in that past place in the present. I call this the Past-Present. You are walking backward into your future, facing the past.

How does a life of imagination differ? First, consider the potential power of imagination. Everything that has ever been created existed first in imagination. This book is not necessarily for just those with a belief in God, but from my faith tradition, even creation at the beginning of the world started with God's imagination. The ancient Hebrew scriptures reveal that God created people *in His image*. "Image" is the root word for imagination. God imagined people and created them.

Need something more tangible? Okay. Pick any object in the room where you are currently reading this page. What do you see? Focus on it. Now, consider how it came to be. Someone somewhere imagined it. They conceived it in their mind. It wasn't yet a reality. It existed only in imagination. Then, the designer, the engineer, or the craftsperson created a new reality. Something that had previously not existed came into being!

The vision of a preferred future—something that isn't yet—is the beginning of bringing that future into being. In this way, we create the future we believe in. Again, from my faith tradition, this is a God-given ability. Creation is God-like. We aren't God, but we have this godly attribute as creative beings.

Here's the big up-side. This can bring us into *Future = Present* living.

Future = Present living allows us to experience the future now before it is reality.
Remember, living from logic and memory, living in the emotions of past events is called Past = Present living, where we live the past every day in the present.
Imagination allows us to live in the Future = Present.

Can you imagine learning to begin living *your future right now*—the one you imagine—in this present moment? This transformational way of life is often called "manifesting." In religious traditions, it is referred to as living by faith.

Here's how this played out in my personal *Bounce* experience, and this is an example of how you can experience this in your life.

I saw in my mind a church unlike any church I had ever been part of. Our specialty would be imperfect people. Let's be honest, everyone is imperfect, so this sounds like an over-broad generalization, but our church would specialize in those who knew it, admitted it, and were working toward transformation.

I could "see" our small team investing in the lives of those who knew they were made for more but needed to learn how to get out of their own way. We would be a church for people broken by their circumstances, their choices, or the choices of others. I not only saw but allowed myself to feel what that church experience would be—and I embraced those thoughts and emotions.

We would be a safe place to explore faith, while being authentic. No one would need to pretend to be better than they were, smarter than they were, spiritually further along than they were. No one would need to pretend to be in any way different than they were. We would be an inclusive community where all would be welcome, and everything would be possible.

I saw it. I believed it. I felt it. I began to live, knowing it already was. This is the mindset critical for successful Bouncing!

My imagination was filled with this vision. Now, let's connect the earlier truth. I had to separate out my own experiences of church and my past pain related to my own journey. I could not allow the *Past = Present* to intrude in any way on the *Future = Present.* That would be like a nail from the past puncturing the balloon of my future before it could *Bounce.*

I had to guard my mind.

This is a point that warrants repetition. It's the part of this process that most people overlook or underestimate. I allowed my emotions to connect to my imagination. It's not enough to just see a preferred future. This is why vision boards or Post-it notes with goals or affirmations are not enough. I had to deeply believe in what I created by imagination, so much so that I could emotionally feel what it was like to walk into such an accepting, safe place—before the place was reality.

I allowed myself to both *believe and feel* the reality of meeting men and women who were on their own journey of exploration and seeking truth, finding their way to their own divine appointments and destinies. I had to feel the safety I imagined, feel the energy in the building as people gathered. And this had to happen *before* any of it *would be.*

This is where I believe most people fail in creating a life of resilience, power, and abundance. They see what they want. They engage their mind, but they don't engage their emotions. Don't forget this: Emotions are more powerful than logic.

Here's an example from the world of advertising. If an ad—on the internet, for instance—merely informs you, that's an appeal to logic. You'll soon forget that ad and will be unlikely to make a purchase. However, if you *feel* something from the ad, you are very likely to explore the purchase further.

This is the process of emotional attachment and imagination.

> ### *You need to believe so deeply in the future that you both see and feel it.*

Now, the process of bringing it to be begins. This is exciting. This is manifestation. As you watch what once was merely a notion, then a crystallized idea, then an image, then a deep belief, then an emotional set-point, and then you add to it the power of words and inspired actions, you begin to bring this into reality. You share the vision with others, and you articulate, clarify, and accurately express what had once been only inside you. Now, you are breathing out life-giving words.

We all know the power of the spoken word. How many still live with words spoken years ago that continue to have a profound impact on their life today? The vision must be spoken. The dream must be put into words. We must speak it. In the same way that words have the power to diminish us, they have the equal and opposite power to bring things to life.

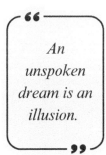

An unspoken dream is an illusion.

An unspoken dream is an illusion.

We know this from the fields of performance coaching and personal development. People are coached to speak kindly to themselves, to speak about themselves based on their beliefs, emotions, and now their words. The most important words we speak each day are the words we say *to* ourselves *about* ourselves.

Now, let's enlarge this concept to your future. Your future is unlimited to the extent you have the faith (belief/imagination) and necessary emotional set-points, and you speak about it like an achieved outcome before it becomes a tangible, empirical reality.

I met with individuals and teams of volunteers, spending countless hours pouring from my mind the church I could see. And I shared it with specific words that reinforced my belief and theirs. I then added to it the emotional impact of the vision. As others saw it, felt it, and spoke about it, our frequencies rose and combined into a palpable energy. This enthusiasm attracted others to the vision. It grew and grew before it ever was.

Conceptually, I have presented how my *Bounce Back, Bounce Better* started. There is much more to this story. But before we pursue that any further, let's get a little deeper into the roadmap for the journey.

CHAPTER SEVEN

Old Dogs and New Tricks

Starting a business (or church) from scratch is much like building an airplane while in flight.

There is a bunch of guesswork, scrambling to make it happen, trying to keep it flying, and white-knuckling the details. I fell back on a practice I learned as a young athlete in school. I went looking for the best coaches in the business. I cannot overstate the value of having a mentor or coach in areas of your life where you desire growth. You will save yourself so much time, pain, and money. Even if you need to spend money on a coach, the net return is far greater in the long run.

I had the opportunity to fly from sunny Southern California to Minnesota in the winter (*brrrrrr*) to learn from experts how to start organizations like I was starting. I listened to as much audio information as I could consume and read books voraciously.

And as I write these words at age 63, I am in the process of pivoting to corporate speaking, book writing, digital content production, and personal development coaching. I've been doing such coaching for years, but for free. I'm engaged in classes on business, marketing, and sales in a technologically based online space. I'm even deep-diving to learn all I can about AI (Artificial Intelligence).

To that end, I have chased, hounded, and written to the experts to get an hour of their time. I've read every book of each person who is best in these burgeoning fields. I've listened to hundreds of hours of podcasts, I've taken paid online courses, and I've consulted with successful people to discern their wisdom. The cost of my education in this new pivot has run in the thousands. The return will be hundreds of thousands. Always be open to this trade. And the result is a master's level education. I am now an influential voice on "mindset" for my clients and thousands more on social media

Here's what I know: You can be mentored by anyone whose life and career you want to follow because of the vast abundance of content these women and men have produced. You can hear from them in their own words, read their stories, and inhale their energy and wisdom, mainly from the comfort of your own home.

The blueprint is available. The talent and ability to be resilient, powerful, and abundant is yours already. It's already in you. Now comes the hard work—tapping into it and expanding it.

Setting up a church in a theater required buying a truck and a trailer for everything the church owned: sound system, kids' crafts and resources, light kit, and promotional banners. Volunteers literally unpacked everything from a trailer each Sunday and repacked it at the end of the day. Every Sunday. For years. Our set-up team bonded, and their joyful service each week inspired me.

And I was right there encouraging, cheerleading, and carrying in equipment. Each Sunday, I started my day at a nearby Starbucks, ordering drinks for every volunteer each week. Our set-up/tear-down team was the best of the best: good-hearted people doing the necessary work to keep this plane flying while we still scratched our heads at times, figuring out best practices and the next right steps. I'll never forget the smell of popcorn wafting through the room as our services ended and the theater geared up to show films.

As the first year wound down, we averaged just over 200 people regularly attending each weekend. We further supported the messaging through our website and social media accounts. It was not the 20,000 to whom I used to speak, but it was satisfying, and I remained very grateful for the opportunity and grace that allowed me to engage once again in such meaningful work. I bounced back. Later, I would understand how I was beginning to bounce better.

At the end of our first year, the theater management informed us that we would have to find a new home: The higher-ups had decided to start showing films earlier in the morning so they wouldn't have the gap for us. The search started. The leadership team and I looked online and in person at well over fifty locations. None worked: price, location, parking issues, or other variables knocked out each one.

With the clock ticking and no new home, we finally landed in a hotel ballroom. The room was large and beautiful and met our needs wonderfully. Sort of. The one part of the arrangement concerned our kids' program, which ran concurrently with our adult program.

The only space available for Kids' Ministry was a hotel room, where each week, we had to move out all the furnishings, set it up for our kids' program, and do it all over the next week. Further, the rooms were down a hall, past the hotel bar. I used to joke that we were the only church in town where kids had to walk past a bar to attend Sunday School. "But very few of them ever stopped for a drink," I'd add.

That started a five-year run at the hotel of setting up and taking down, not having weekday space for activities or meetings, and knowing this wasn't our final home. During our time at the hotel, management would sometimes book big events that bumped us from the space. We'd scramble madly to find a temporary location for those weekends. We did this frequently enough that I said, "It takes intelligence and ingenuity to attend our church, considering how often you have to find us."

After five years at the hotel, we decided we had grown to the point where we needed more permanent space and weekday space to scale the work we were doing in our community and for our region. The search began again.

And again, dozens and dozens of options were explored. Finally, the largest space in a retail center became available. With a renovation, it started serving our needs wonderfully. The rent was fifty percent of our monthly income, putting a strain on our programs, but we took the risk and experienced a new season of growth as a result.

Are you applying what you're learning to your life circumstances? You have all the "potential energy" that is the basis for *Bounce* in you already. Divine provision is on the way and scheduled to arrive right on time. Your future can be substantially greater than your past. And there will be a season of hard work and many pivots along the way as you take inspired action toward the future you imagine and feel; your future = present.

OLD DOGS, OLD VOICES, AND NEW TRICKS

Through these years, I remained very grateful for the opportunity to *Bounce*. Critics remained; we were engaged in a constant juggling act to deal with financial pressure. The challenges of growing a new venture were consistent, and then we were hit with a global pandemic. Yet, I saw each of these challenges as strengthening my *Bounce* so I would be prepared for even higher heights.

You know the story of small- to mid-sized businesses during that dark period in the world's history with the pandemic. In Los Angeles County, we were forbidden to congregate in person. That could have been the end of us, but something happened just before the pandemic that changed everything.

We made a strategic decision to increase our reach to an online audience. Quickly, viewers from many states and even several other countries became part of our weekend programming through the internet. We invested in the technology to make this a terrific experience. So, when the pandemic hit, we were able to pivot. We added podcasts and video archives to broaden our technological reach. Soon, we were reaching well over a thousand people a weekend *during a pandemic!*

The need to always innovate, to always explore and take a risk on the new, is vital to staying ahead of unknown forces that clamp down on profit, sales, success, and other issues you might face in your business or life. Technology saved us.

In addition to technology, I made the decision to move our weekend church services out to the parking lot, where we were allowed to gather. Thankfully, it was in Southern California weather. For nearly a year, we gathered on folding chairs, and from a portable stage we conducted some of the most fun and creative experiences in our history.

Innovation, technology, and a pandemic came at me fast. As someone with nearly four decades of doing something familiar, this was entirely unfamiliar. I was not tugged from my comfort zone; I was *ripped* from it.

> *Old dogs can, in fact, learn new tricks.*

What I know from those experiences is that no one is too old (or too young) to experience new and wonderful parts of life, to discover depths about themselves, and to BOUNCE in directions that are invigorating and stretching.

Old dogs can, in fact, learn new tricks.

Old voices were also a factor. With resources thin and growth gnawed at by the pandemic, I felt my thoughts reverting to old tracks, old tapes, and old voices. I began to think, *I'm not enough. I've gone as far as I can. I'm too old, too limited, and God has probably blessed me with all He is going to.* Many people have shared with me how this happens in their lives, too.

Here's why this is an ongoing issue that all of us must wrestle with on the road to resilience, power, and abundance: Our minds automatically go to the familiar. This is our default. And we have spent years listening to limiting thoughts, the repetition of "not-enough-ness" filling our waking hours. Is it any wonder our minds will drift to those well-worn ruts?

The problem with a rut is that it gets deeper over time, eventually becoming a grave.

The constant vigilance required to keep your head in the game, to think better thoughts, and to hold fast to the right beliefs and mindset is the work of any successful person. I had a coach who used to say, "Keep your head in the game." I'd put it like this:

***Keep your head, heart, feelings, words,
and actions in alignment.***

It was vital for me to be able to recognize when old tapes would play so I could reach for the volume control in my mind and silence those thoughts immediately. But neurologically, we know that silencing an old tape isn't enough. You must replace it.

The meditative experience of waking every day, setting intention, and buttressing that with new beliefs, emotions, words, and actions is a secret that great dream-achievers know. Now, you know it, too.

One more event transpired during the pandemic that altered the course of my life and thinking.

Through the struggle of my early childhood and formative years, one person was my rock: my brother Jeff. He was my constant source of assurance that, somehow, I would be okay. Two years older than me, Jeff was my very best friend, and our love was unbreakable. This is the brother, my only sibling, who was on the other end of the phone when I woke from my third suicide attempt. He had been there for me every single day of my life.

Then, Jeff contracted COVID-19. On the day after his 63rd birthday (my current age as I write this book), Jeff passed from Earth to heaven. With both our parents already gone, my family of origin ended. And with his loss, I questioned everything, including how we are to live our one and only life—a life that at its very longest point is very short. I will write about this later in the book.

There are days I still grieve, and I miss him terribly. Part of me died with him. And yet, part of me came alive, and I'm determined to live my best and fullest life because Jeff would want that for me and that's how he lived his life.

That set me up for exploration about what's next, what's ahead for me, and what I believe I am to accomplish. With fresh clarity about my purpose, I now live my life with a burning passion and urgency to become, experience, and accomplish all that I am created for.

And that kind of full, rich, happy, fulfilled, meaningful, giving life is available for you, too! This is the season where *bouncing back* can become *bouncing better*.

CHAPTER EIGHT

Dreaming a New Dream

Here's a fun exercise to put your mind through: If you could do anything with integrity and purity of heart, anything at all with your one and only life, and know you wouldn't fail, what would you do?

I've used this exercise with thousands of people, and they generally have an answer. And what is the number one reason people give when asked why they haven't pursued the one thing they just said they truly want to do? The answer is baked into the question: FEAR. Fear of failure, fear of judgment, and, for some, even fear of success.

So, let's sit with each of these fears, one at a time. First, consider the fear of failure. This is the most interesting one to me because of what it implies about the definition of failure. The implication is that if I try and don't succeed, the only other possible alternative is failure. But what if we aren't operating in a binary (only winners or losers; successes or failures) or linear (I must track upward on a graph to succeed) world?

Consider this alternative way of thinking:

The only true failure is the failure to try.

See, once you try, regardless of the outcome, you will have more data points and you will be closer to learning what *will* work. So, when you try again, you try from an elevated posture; you've already risen—even a little bit—and have mathematically already increased your likelihood of achieving what it is you want to achieve. This way of interpreting failure is to reframe it as learning. This is commonly called a "growth mindset." All wildly successful people have this mindset.

Like many others before me, I assumed someone won or lost, succeeded or failed. Now you know a better mindset, and the threat of fear is disarmed. Failure doesn't exist for the person who tries. (We'll explore fear in more detail in a later chapter.)

My fear, however, wasn't the fear of failure. It has always been the fear of judgment. Because of my very public humiliation and subsequent criticisms for all the harm I did to others, I grew exceedingly fearful of ever raising my head out of the foxhole again. You may recall something I wrote earlier: My plan was to go through life with my head down.

When you attempt to change your life, to do something new, to take on new levels—you are also taking on new devils. Critics will sharpen their knives and troll your new website, cast vitriol into your new social media sites, and do their best (don't miss this) *to keep you from rising above their level.*

See, those above you seldom cast aspersions on those climbing on the trail behind them. It's those below you who will seek to bring you down.

This is a sad, dark side to a social media and internet presence. No matter what you do, someone will criticize what you're doing or why you're doing it, believing they know better what you should do with your one and only life. My friend reminded me, "Brad, even if you could walk on water, you would be criticized for not being able to swim."

I'm calling "bovine excrement" on that.

Do not allow the fear of criticism or judgment to keep you from achieving the fullest, most complete, most extraordinary life you can live. Yet here I am, confessing that, for more than a decade, I was held back from building the amazing life I could have by this singular fear of getting back into the arena, raising my profile to a more public place on social media, and stepping out of the expected role others had for my life. Those are years I cannot reclaim. However, I am now making the most of the years remaining. And I can help others avoid the mistakes I made myself.

Here's a compelling question: When did you decide to give up on your life?

When exactly was it that you settled, conformed, stopped growing, and began to watch the clock run down on your future? I pray that your answer is: I haven't given up on my life or future. Or, like me, some may say: I had given up, but no longer. I'm getting back in the game!

I promise, I promise, I promise that you will dislike the result of not attempting greatly *more* than you will dislike the criticism that comes when you step out of your comfort zone, step out of what others expect from you, and make giant strides toward the life you've always dreamed of living.

Then, some fear success. They aren't afraid of all aspects of success: positive influence on others, financial freedom, and inner satisfaction from pure, noble work. What they fear is what success might do to their daily schedule, the demands on their time, the dynamics of close relationships, the travel that might be involved, and the stresses of keeping momentum going in something they build.

Let's just take a momentary deep dive into the financial side of this. What is more stressful: a busy, full schedule or being broke? Busy or broke, which do you want?

Yes, success of any type—academics, athletics, relationships, business—has hard parts. So does poverty. So does a life without excitement or meaning, a life with little to no impact, a life without close relationships.

Choose your hard.

Choose your hard.

If any of the above three fears resonate, make a list of all you want to achieve in your life on one side of a piece of paper. On the other side, list all the possible downsides to trying. Then, pick the side of the paper you want to live on: attempting or excuse-making, daring greatly, or shrinking into the hard parts you have already.

When it's stated in such obvious ways, the answer for you will become obvious. None of us want fear to win. And all of us *can* bounce above the fear to the life we've always dreamed.

A SERENDIPITOUS CONVERSATION

For me, I have deeply fulfilling work. I get to impact people, see lives change, and truly serve my fellow sojourners on this planet. From the outside, this is remarkable.

What others didn't see or know is that, while I did all that, there was stomach acid that churned inside of me when I couldn't pay all my bills, when I checked my bank account multiple times a day, knowing we were living on the brink of being overdrawn. Always, two days before payday, our entire financial lives hovered somewhere around twenty dollars to one hundred dollars to our name. Every two weeks, I held my breath and hoped nothing would crash, break, or demand our final few dollars.

The civil war between mind and spirit went something like this:

Spirit: *But sacrifice is part of serving God by serving others.*
Mind: *Yes, but you have to pay your bills.*
Spirit: *Of course I do, but God will provide.*
Mind: *What if He wants to provide through hard work that pays more?*

The conversation ran on a loop in my brain, mainly between 1:00 a.m. and 3:00 a.m.

Something had to give. When what you do repeatedly brings the same monthly belt-tightening, it's time to find a new strategy. Before my utter collapse in every realm of my life, I had earned an extraordinary living. I had stress in my life, but not financial stress.

This new reality of financial stress every week—month after month—began breaking me down inside. Some people appear to roll with financial stress. I am not one of those people. Having a financial base that's strong is important to me, for my family, and for our future. Yet, something needed to change for that to happen.

While I was running on emotional fumes due to financial stress, I had a burning desire to multiply the number of people I could impact. I had grown content with the sweet congregation I served and could have continued to serve them, kept my head down (posture of shame), avoided critics (a recurring fear), and whiled away my one and only life. But I knew I was made for more.

As I healed, I grew more confident and determined to make the most of every minute I have on Earth. My thoughts increasingly turned to a move from retail to wholesale.

Imagine a mom-and-pop shop. It's a small, intimate family business with extreme customer loyalty. But like all mom-and-pop shops, customer loyalty isn't the crushing issue. It's the size of the customer base. But in our imaginary story, Mom and Pop toil along each month, nail-biting as they go through the books, and they just hope they will survive another month.

Their life isn't all bad. Again, they have a loyal customer base, and they love serving their local community in the ways they can. Yet, they are withering inside because the financial stress weighs heavily on their mind. It's for this reason that the majority of mom-and-pop shops shutter.

Another approach is to move from that retail shop to wholesale, where the business is supplying all the shops and retailers on the front lines. Now, rather than having the customer base of one shop, you have the chance to impact at 10x or even 100x.

I once had the privilege of serving and helping to effect life-change for tens of thousands. The desire to do that again was coming alive within me. I felt a pull from retail to wholesale. For the first time in years, I began to believe that I could not only *Bounce Back* but *Bounce Better*.

This is when a serendipitous conversation occurred. Family friends invited Karen and me to dinner. We have much in common with this couple and enjoyed our time together at their beautiful beachfront house. On one specific visit, we strolled along the ocean, had drinks in Adirondack chairs in the sand, and savored a delicious meal with a water view.

The casual conversation turned to deeper, meaningful topics, like how the talks I prepared week after week, they said, had the potential to help more people. This couple asked if I had considered expanding my reach and touching thousands of lives with my content. My gut knotted at what that would mean for me; I'd have to lift my head up and walk into my community with more exposure. It all felt very risky. Yet, that precise gnawing thought had been growing inside of me.

Who could know that the seeds of their words (and their belief in me) would germinate over the next few months into a complete re-arrangement of my life—in a wonderfully compelling and adventurous way?

A REINFORCING PATTERN OF THOUGHT

Have you noticed that once you've had a thought that's meant for you, things happen to reinforce it and drive it deeper into your heart? Some people call this the Law of Attraction. To me, it's a keen spiritual awareness of all that's aligning around us once we determine what we want.

A short time after our meal with our friends, Karen told me about a podcast she'd listened to that she thought I'd enjoy. I did, in fact, enjoy it and found myself enthused, as a very similar message emerged of taking one's gifts and skills and turning those into something that helps others and *generates* a strong income for family and future.

Karen and I listened to other similarly aligned podcasts, and we then ordered and read a myriad of books. We held late-night talks and early-morning dream sessions together, considering what life might look like if we put shape and form on dreams that now filled our conversations and waking hours. I subscribed to paid coaching sites and began taking pages of notes. I immersed myself in learning everything I could on how to reach a larger and larger audience, using content I created that I knew could help improve their lives.

But to do what we were considering would no doubt bring out detractors and haters, and potentially, even friends may not understand why we would attempt to do more than we were already doing. There is a downward gravitational pull in groupthink that suggests, *Just keep doing what you're doing. Don't rock the boat. Stay in your lane. We like you just the way you are.*

I had to put to rest the fear of the judgment of others once and for all. If we are true to ourselves, pure in our motives, clear on this new meaningful mission, and walk forward with integrity then no *one else's voice* should matter.

Here's the alternative (this applies to you, also): Who will I allow to determine my future? Critics? Those not bold enough to step into the arena? Those who benefit from my life not changing? I think it was podcaster Mel Robbins who said to be very careful who you hand the microphone to. What she meant by that is that you need to carefully choose those who you allow to speak into your life and your future.

Karen and I worked hard to ensure our intentions were clear and our hearts were pure. We studied and prayed and then studied and prayed some more. We understood that a movement of life and spirit was beginning within us that would have far-reaching implications for the rest of our lives. This was an epic time of transition for us to step into not just the bounce–we had already bounced back–but to allow ourselves to believe and aspire to the *Bounce Better* part of life that we now firmly knew was possible for us (and it's possible for you!)

With this understanding, we went to work building the mindset to support our next level.

CHAPTER NINE

Mindset Matters

Earlier in the book, I said I had left my earlier commitments and former life long before I took any discernible steps to depart physically. Leaving begins in the mind. This same principle works in reverse. It became clear in my mind what my next season of life would include. I envisioned all the details, emotionally attached to images, and spent time feeling myself living the life that now had clarity in my imagination.

I saw myself living in this new life. I could close my eyes and feel what it was like to wake up each morning, ready to enter my day. I imagined my morning rituals; I thought the thoughts I'd have as that person in that place. I smelled the citrus blossoms from the window, tasted the coffee, and felt the confidence of doing noble work. I knew the clothes I would wear, the car I would drive, and what my workspace looked like in my life, which already existed in my mind and in my future.

This (and more) all congealed in my mind until my emotions created a brain-chemical reaction, which happens with all strong emotions, so now I had both an emotional bond and a physical bond to the future. This allowed me to create the future, to experience it emotionally and physically in my present well before it came into being. This is an important part of *Bounce* and one that you can do too.

The height of your Bounce is determined by what your mind imagines and what you truly believe you deserve.

SELF-SABOTAGE

Looking back on my previous climb and fall—rising star to falling star—I had to consider something a psychiatrist said while adjusting my depression and anxiety medicine during the depths of my broken life. I don't recall his exact words, but the message was this: *You soared higher than your self-esteem could accept or believe. So rather than waiting for a "crash and burn" to happen, which you were sure it would because you felt like an imposter, you decided to take matters into your own hands and sabotage the whole thing.*

Wow. Self-sabotage is actually an effort to stay in control of the outcome, to control the narrative of one's life. It's counterproductive and harmful to oneself and often to others, but it does allow one to stay in control. It's concluding at a subconscious level: I will control the end game. I will control the trajectory. I will not wait for what I believe will be a fall, or failure, or reversal of fortune, or embarrassment of being exposed as 'not enough.'.

If I was capable of such sabotage once, I wondered if that would be the inevitable place I'd find myself again as I began to *Bounce Better*.

This is a very real phenomenon. We've all heard people say things like, "I broke up with her before she broke up with me," or "I quit that stupid job before they laid me off." People self-sabotage, consciously or subconsciously.

In an episode of *America's Got Talent*, Karen and I once saw a contestant quit prior to hearing if she'd made it to the next round. When asked why, she said, "I probably wasn't going through to the next round, anyway."

Do you see what's really behind those statements? It is more than fear. It's a grab for control. Rather than allowing something bad to happen to me and having to take it passively, I'll short-circuit the process and just pull the "bad" lever myself. Crucially, such choices are all predicated on the belief that it's going to turn out badly anyway. *That* is self-sabotage waiting to happen.

Here's an alternative mindset: What if it turns out better than you thought? Allowing ourselves to get past limited thinking and "not-enoughness" thinking requires intentionally focusing on new thoughts. It must be intentional. This is why we call this part of mindset adjustment *meditation*. Meditation is nothing more than focused thought. If we believe (see that thought, hold that thought, feel that thought) good can happen in our lives, then we won't short-circuit the process. We won't break up first. We won't quit the job or the competition, or we won't burn the whole thing down with destructive choices (I was Exhibit A on that one).

Assessing old patterns of thinking that work against you is what I mean when I say, "We must get out of our own way."

But these were just a couple of the mental hurdles I had to clear on my better bounce.

CLEARING AWAY MISCONCEPTIONS

Here's a truth that was life-changing for me:

You cannot manifest the life you imagine if you have contradictory beliefs that block your future from becoming reality. Our achievement will only rise to the level of our beliefs being able to support that achievement.

If we say we want "Dream A" but inwardly hold beliefs that tell us we either don't deserve "Dream A," or aren't enough to achieve "Dream A," or believe that "Dream A" is morally bad, no matter how much we meditate, no matter how many dream boards we create, and no matter how hard we work, we will never achieve higher than our belief system.

Are you seeing this again and again on these pages? It is about beliefs and mindset. Every day. All day.

When Karen and I sat down and began to have honest and complete conversations about our financial picture, old, lurking thoughts rose to the surface of my mind, and I realized right then that *these* are what kept me from earning more all these years:

- A belief that I should work for others rather than for myself
- A belief that wanting more was selfish
- A belief that desiring to earn more was bad. (Aren't Christians supposed to be humble and poor, after all? Let's save that one for another book.)

And then a big one:

- A belief that people would judge and criticize me if earning more was my stated goal

Let's take these on one at a time. Perhaps you'll recognize some of these same limiting / blocking thoughts are in your mind stifling your *Bounce*.

Misconception Number One: I should work for others and not myself.

Allowing the organization you work for to set the cap on your overall capacity for life, the positive influence you can have, or the amount of money you can earn is a choice. It's impossible to support our decision to earn more or the belief that we are made for more if we determine we aren't entrepreneurs and that we don't want to work for ourselves. Another possible issue is that we fear we aren't capable.

The counter-argument is that if you work for yourself, you are empowered to pursue every goal and every dream and *BOUNCE* to any height you desire. It's invigorating. It's freedom.

I had to move from "I'm just an employee who will be told what my capacity is" to "someone who takes responsibility for my future and is excited to do so."

Misconception Number Two: A belief that wanting more is selfish.

I was listening to personal development expert Jen Gottlieb's podcast and she said her first step to changing her life in a terrific way came the day she admitted that she wanted financial freedom. She then said, "It used to feel like a selfish goal."

I know it can be unless you reframe that.

Imagine this: Every charity needs donors. Can you see yourself being a major donor? Do you have elderly parents who currently need care (or one day will) that you could pay for? Life is about loving God and loving people. If that is the basis for me wanting more, then, with a pure heart, I can move forward without the attendant guilt.

Misconception Number Three: A desire to want to earn more is morally bad.

I have vivid memories from my childhood of Mom and Dad telling my brother and me not to tell anyone when we got a new car because members of the church congregation would think they were paying my dad too much as their pastor. Dad and Mom would drive our old car to church and keep the new one hidden in the garage. Or we couldn't talk about a vacation we were taking for the same reasons. Or we could only shop at Sears because that was acceptable for a middle-class family. Such behavior left me believing that having money is somehow bad and things like purchases or vacations will invite negativity.

This isn't likely as true in many professional circles, but it was one of the misconceptions in my experience. And like a wet blanket, this belief structure layered over my hopes and dreams.

Verses from the Bible were used to make wealth seem evil:

"The love of money is the root of all evil" (notice, it didn't say money).

Or this one

"It's easier for a camel to get through the eye of a needle than for a rich man to enter the kingdom of heaven."

When you grow up believing wealth might keep you from heaven, that's a serious wet blanket weighing down your dream balloon.

I don't want to get preachy, but allow me a moment to reflect.

It would certainly be unwise to ignore the built-in caution from these scriptures. If one loves money more than people, let's say we could all agree that it may be a condition of heart and of priorities that limit our best self. If one, however, loves freedom, the removal of the weight of debt and financial stress, then that becomes an entirely noble pursuit.

I've known many wealthy people whose heart for God and others is a shining example of all I aspire to be. They are generous, wise, humble, and kind. We have also all seen examples like Dickens' character, Scrooge, who is miserly (same root word as "miserable"), tight-fisted, and uncaring. These are two different approaches to wealth.

If we allow these verses to keep us poor, financially struggling and stressed, we have misapplied the deeper truths. Both verses (and others), when taken in context, always, always, always refer to the grip possessions have on our heart or the place that money alone occupies in our life.

There is a woman named Lydia mentioned in the Christian scriptures. She's called the "seller of purple." Purple dye was exclusively the color of royalty and Lydia was a businesswoman who sold the dye. Her clientele was elite and wealthy. Lydia was therefore successful and wealthy. We also know that Lydia helped finance the ministry of a very famous missionary and writer, the Apostle Paul. Wealth was obviously a blessing.

Having led religious non-profits for the better part of forty years, I know the value of generous donors and have seen the wonderful impact of money given to noble causes. Do we really think being an economically strapped CEO or pastor or president of a non-profit is somehow righteous, while that same CEO, pastor, or president gladly accepts donations from wealthy supporters who are somehow morally bad because they have achieved success?

That would be both hypocritical and foolish.

Achieving wealth and being wise, generous, and humble with wealth is honorable and, I would argue, preferable to poverty or financial stress. I had to wrestle with this until I could get my mind free from the grip of the misconception that having money is bad.

Misconception Number Four: A belief that people would judge and criticize me if having money or success were stated goals.

It's not untrue that there will be trolls and there will be haters anytime we step up and step forward to pursue the life God has planted in our hearts. However, is that reason enough to abdicate our own life to silence the critics and give up on our dream to appease the haters?

That would be dumb, dumb, dumb.

Why? Cause haters gonna hate.

The minute you adjust your life to fit the opinions of others, you have given up your power to create the life you want. And you have given power over your life to the very people who are haters.

Then, you are living the life someone else wants you to live. And they have opinions about your life for so many reasons.

I'm sure some people looked skeptically at Jamie Kern Lima, the billionaire founder and CEO of IT Cosmetics, when she was waitressing at Denny's and dreaming of becoming a broadcaster. Some likely guffawed at her dream. They thought of her only in terms of waitressing. Waitressing is honorable. There is great virtue in the hard work required in that profession to serve others. Remember, I was a barista. But Jamie had a different dream. And that is okay, too.

There will always be those who want you to stay in your lane. They have perceived you one way and cannot make room in their mind for you to be anything else.

Some will criticize you because your courage to change shines a bright light on their complacency. Like it or not, the minute your ball starts to bounce and rise, others will feel a personal threat and hope your ball bursts.

Even more common, when you chase your dreams, are negative responses from people who are currently served by you being who you are and doing what you do. They will try to talk you into staying just the way you are because it serves their needs. This is a back-handed compliment because they value the role you play in their life, and they don't want to lose that. But again, that's selfish on their part, though they'd hardly recognize their attitude that way.

Criticism in our culture is a sad reality. Expect it, but don't let it define you, and don't let it stop you. If that criticism is coming from someone you wouldn't ask for advice, or if it's coming from someone whose life you don't want to imitate, then why would you let that enter your mind and defeat your spirit? And even if it comes from someone close to you, if your heart is pure, your dream is noble, and it's burning brightly inside you, don't let anyone extinguish that light.

Very few above you will criticize your desire to *Bounce Back* and *Bounce Better*. Mainly, it will come from those below you. Keep that perspective in mind when criticism hits.

Let's take a deeper dive into these topics.

CHAPTER TEN

The Death of How I Thought It Would Be

Your imagination, mind, emotions, beliefs, and actions have worked to create your future. It exists in the place called faith. It's seen before there's empirical evidence that it exists. You have set this intention, and it's already real in you. You see it, feel it, and live in it. You are now making moves toward it as the new person you already are. It's not that it *will* be—it's not that *you* will be. *It is already,* and *you are already.* You are steadily accessing what is.

In order to get this solidified within you, so you can then move from intention to manifesting—bringing it into reality through aligned beliefs, emotions, words, and action—you must pass through what I call the "Four Deaths." The greatest life *that's already real* is accessed through these deaths. In my faith tradition, we might say the new life of resurrection begins after the grave.

Passing through the four deaths reframes everything about your perspective. These became four cornerstones and the foundation for your brand-new life and level: your *Bounce Better.* Let's jump into the journey of the Four Deaths.

Let me start this section with two stories.

A woman went into her marriage with romantic feelings and childhood dreams of how she thought her marriage would be. In her view of marriage, kindness prevailed, hard work led to financial stability, and each person in her family embodied health and optimism.

But the week before her wedding, she was diagnosed with a crippling disease. Her fiancé married her but soon tired of the overwhelming medical bills and her inability to partner with him in ways he needed. He drank more, and his moods and behavior grew angry. Soon, because he couldn't stay sober, he lost his job and lost the health insurance they desperately depended on for medical care.

Lying beside each other, listening to the frustrating huffs of the other's stressful breathing, they both had a thought at the very same time: *This isn't how we thought it would be.*

*****

A university student locked himself in his dorm room or at the library every night and weekend to study. His friends partied and took their coursework casually, but this young man wanted more and believed he could have more if he applied himself.

He graduated very near the top of his class, and his prospects for the future were high. He entered the job market, worked with recruiters, and landed a position in a firm that he believed gave him great opportunities and prospects.

Because the company only employed top graduates, he was no longer the lone shining star of his class but just one of many talented and ambitious young people at this firm. As a matter of fact, a couple of the others were not only bright but also savvy with office politics, worked their way into favor with hiring teams, and quickly rose in the ranks.

This young man languished on the bottom tiers of the company, occasionally thankful for a bonus that came his way, but while others climbed, he felt himself settle. While others dreamed of the life that could be, he became cynical.

One evening after work, he sat alone at a bar. Staring into his drink, he said to no one in particular: *This isn't how I thought it would be.*

*****

Perhaps this has happened to you—in a career, in a relationship, in your health, for your family, or your future. You've found yourself contemplating the discrepancy between how you thought it would be and how it really is.

Did you know that happiness studies have shown us that the happiest people accept what is, while the unhappiest rail against the inequities, complain about the unfairness, and seethe in resentment that life isn't the way they thought it would be?

This happened to me. I was in the top 1% of income earners in the country, married with two children, lived in a million-dollar home in a gated community with a pool in the backyard, and every metric in my professional career tracked up and to the right.

Disappointment is unmet expectations.

But I was miserable because there were parts of my life incongruent with my expectations. That led to a drowning sense of disappointment.

Disappointment is unmet expectations.

I lived disappointed. There were parts of life that were not stacking up the way I thought they would. People weren't responding to me in the ways I had hoped, and my internal needs weren't being satisfied as I expected.

I grew angry. My anger turned inward, becoming depression. It also led to destructive choices, which led to shame, and shame and depression led to my three suicide attempts.

Fortunately, I failed at my attempts at *failure*. Today, I have a semi-colon tattoo on my right wrist, which means that my story isn't over yet.

The good news is that your story isn't over, either. But how do you deal with the disappointment of not being where you thought you would be, not accomplishing what you believed you would, or not having the relationship(s) you hoped for?

This reality of incongruent living—where life isn't what you thought or hoped—has the potential to break you or make you.

You may not have the life you thought, but you can have the life you dream.

Life wasn't what J.K. Rowling thought it would be. The world-famous author of the "Harry Potter" series experienced many setbacks before her meteoric rise to success. She experienced a failed marriage, the loss of her mother, and struggled with depression before finally achieving immense success with her writing.

Life wasn't what a poor African American girl thought it would be. Despite being born into poverty and enduring a difficult childhood that included physical and emotional abuse, Oprah Winfrey found a way to become one of the most influential media personalities in history.

And let's not forget, the happiest place on Earth began inside the mind of a man who experienced many setbacks and rejections in his early career. The legendary animator and creator of the Disney empire was fired from his job as a newspaper cartoonist *for lacking creativity*. He had multiple business failures and saw his first animation studio go bankrupt before eventually founding the Walt Disney Company, which grew into one of the most successful and influential entertainment companies in the world.

How do people like these, and many thousands more, go from not living the life they wanted to creating the life they always dreamed of? They each made the same move: they "pivoted."

You know what it means to pivot, right?

> ***Pivot: Adapt to change, tap new markets, and capitalize on new opportunities***

Though this sounds like an external solution: "*Oh, just be the same person, but do these few things, and life turns around!*" NO!

Pivoting is less about what you *DO* and more about who you *ARE*.

> ***Your future success in life is way more about the hunger in you and way less about the history behind you!***

There are interior characteristics that hold true across the board for everyone who ever succeeded greatly after concluding that their life was different than they imagined it would be. Rather than fighting that, sitting and just complaining that somehow they are a victim of life and therefore stuck, these men and women learned the art of the "Pivot."

THE PRO-LEVEL PIVOT

Allow me to unpack this. Here are the attributes of what I have named the Pro-Level Pivot:

Perseverance

At the very heart of the concept of perseverance is the belief that no matter what, you find a way forward. We are designed for forward momentum.

You know this:
- Our eyes are in the front to help us see where to go.
- Our ears are tilted to the front so we can hear what's ahead.
- Our arms reach forward more easily than backward.
- Our feet face the future.

The only significant body part pointed toward your past isn't who you want to be.

From our anatomy and from the success stories of others, moving forward is the right play.

When life hurts, when life changes, when life isn't what you thought it would be, the answer is in front of you, not behind you. We are designed to *bounce back* and higher, not backward.

In my case, I had to persevere past suicide and deep depression and pivot in order to climb out of the dark place I'd lived in for a couple of years. There are two things that mentally almost stopped my pivot: The belief that I was disqualified because I failed and my fear of the changes I would need to make.

Think about those. Have you ever felt either of them?

A lot of people feel disqualified from starting over. And even people who haven't failed often don't believe in their future success. People disqualify themselves and, therefore, don't even try to pivot for many reasons: I'm too old, too young, not educated, have a broken past, feel unworthy, and no one believes in me. The limiting beliefs go on and on.

Here's what I had to learn: As adults, we are never starting over. We don't ever start all the way back at the beginning. We don't lose the good traits we acquired along the way.

We start over with years of experiences—good and bad, and years of education—most learned the hard way. We always start higher than our first start—despite what our circumstances tell us—because remember: It's not the outside circumstance that is the greatest predictor of future success. It's what's inside of us.

Inside of me was doubt, but I knew that in order to get up and get out of that dark place, a pivot required me to keep going—to persevere.

The other issue was a fear of change. I'd lived two years in the dark place, in the struggle, bombarded with the consequences of repeated bad choices. You'd think I would jump at the chance to change.

In the science of change, we know that 80% of us don't want to change, *even when we know the change will be good.*

Instead, we say, "Well, life isn't what I thought, but I don't want to change, so I'll comfort myself in my discomfort." So, we drink a little too much to *take the edge off.*

We go on a shopping spree and live in denial about the debt. The dopamine from a new purchase comforts us in the discomfort. It's the rush of another Amazon package showing up on our doorstep that eases our discomfort.

But when we want a better life, a different life, we move beyond the discomfort to create forward momentum and persevere despite our past, despite our circumstances and despite our reluctance to change.

It's the difference between pressing the PLAY button on a recorded TV show to see what comes next or pressing the PAUSE button and stopping the action where it is. Most people hit pause, but some of you want to press PLAY, and I suspect that's why you're reading this book today.

At the age of 13, professional surfer Bethany Hamilton lost her left arm in a shark attack. That certainly wasn't how she thought her life would turn out. Despite this life-changing event, she refused to let the incident stop her from pursuing her passion for surfing.

With incredible perseverance and resilience, Hamilton returned to competitive surfing just a year after the attack. She has won multiple competitions and become an inspiration to others, sharing her story through motivational speaking and in her autobiography, *Soul Surfer.*

> *Those who pivot persevere.*

Those who pivot persevere.

Karen and I recently watched an episode of *American Idol*. That show always makes me cry because of the stories of people who overcome such odds and continue to show up.

It was the time of the season when the judges picked the final twenty-four contestants from over 50 remaining. On the night we watched, the number of contestants was being cut by more than half; more than half were going home.

The judges called the contestants in one at a time and told them yes or no, up or down, "You're going through," or "This is the end of the road." In an unusual move, they called two singers in together. The judges explained that the two were so closely matched they were going to need a sing-off.

One of the contestant's facial expressions said it all: *This isn't what I wanted, and it's not what I expected.* Her disappointment was obvious.

Luke Bryan, one of the judges, said something so important. He said, "Hey, as long as you're still singing, you're still in it."

Quit if you want to… or keep singing. Perseverance

Intentionality

What are your values? What is most important in your life?

Most people would list faith—their relationship with God, their creator. Most would include family. Many would say friends, physical health, and financial freedom.

Once we establish our values—what matters to us in life—we have done two things: First, we've established what matters less. When you know what you value, you know what didn't make the cut and is not a driving value. Second, when we know our values, we can evaluate our decisions. Therefore, we can begin living in an intentional way.

For instance, if I say family matters, I can then look at my choices and my daily activity and I can assess: Did any of that help my family, strengthen my family, show love or care or provision for my family?

We run down the list of our values, and we see if we *intentionally* supported those values.

It doesn't matter if life goes sideways or if a rogue wave hits the boat of your life and capsizes it. It doesn't matter that life isn't the way you thought it would be. When you know your values, you can conclude that you still have the choice to intentionally support, engage in, and work toward what is most valuable to you.

Do you see why so much of this is about what is inside you?

If the stock market crashes, I can still love my family. If I find myself unemployed, I can still value my health. If I accumulate debt, I can still value working toward financial freedom.

Value-based intention sets the course from where we are to where we want to be. At least you can identify the next steps.

Consider this: Do you wake up every day and set your intention? Do you know where you want to go each day? Do you know your intention to support your values that day?

To set an intention is to shape your day with the desire of your heart, the values of your life, and the items that will show up on your schedule for that day.

At the granular level, this involves all of your choices. Just imagine you're a sculptor confronted with a large slab of marble. The art that will emerge will move from your imagination to reality.

But how? One chip of marble at a time. One choice of where you'll lay the chisel's sharp edge at a time; one choice of how hard you'll tap the hammer on the chisel at a time. Those tiny chips and choices will lead you to a masterpiece.

The same is true in your life.

Having completed what would in time become one of the most celebrated sculptures in the world, a depiction of an angel, Michelangelo was asked how he'd done it. He replied,

"I saw the angel in the marble and carved until I set him free."
— **Michelangelo**

Can you see the intention? And do you live with such intention?

Those who do not pivot have concluded they are done chiseling. They say things like: "That's all. Not going to release the angel. Not going to work toward what I want for my family, my faith, my finances, or my future."

But others want a masterpiece. And given that you're reading this book, I think you decided that what you want for your future is a masterpiece.

Versatility

Remember our definition of Pivot: Adapt to change, tap new markets, and capitalize on new opportunities.

Those who successfully pivot when they find themselves somewhere other than where they expected or wanted to be ask one question:

"What is the opportunity in this circumstance?"

It's like the guy who sold plain light bulbs, then adapted to market changes and sold LED bulbs, and then adapted to further market changes and sold lasers. When asked what his secret was, he said, "I realized I wasn't in the light bulb business. I was in the business of light. I wasn't stuck or limited to just selling light bulbs. When the market changed, I adapted to LED and then adapted to laser. And it was easy because my core business was larger than one product or method."

Wow! Did you see that?

… my core business was larger than one product or method.

As I assess my own life, my core business is creating important content and communicating that content so that it changes and improves lives. I can deliver that in a hundred different ways. I can be versatile in how I accomplish my core business.

What's your core business? What is your core life goal? Versatility happens when you determine your big-picture objectives and then, over time, pivot on the who, what, where and when.

Warren Buffet had the objective of being the leader of a profitable enterprise, but he was horribly shy and felt terribly incapable of being a public speaker, as someone who would have to represent his company. So, he invested over a year learning to be a public speaker, and he said that was the single biggest contributor to his success through the decades.

His willingness to adapt, grow and learn new skills propelled him to great success. That's called versatility. It's also called a growth mindset, as we discussed earlier.

Consider these examples:

Netflix began as a DVD rental-by-mail service in 1997. As technology and consumer preferences evolved, Netflix saw the potential for streaming video and launched its streaming service in 2007. This strategic pivot allowed the company not only to survive but also to thrive and become the global streaming giant it is today, with original content production and millions of subscribers worldwide.

Most of you know the brand Nintendo. What you might not know is how the company proved its versatility and pivoted. Nintendo began as a playing card manufacturer in the late 19th century. Over time, the company ventured into different industries, such as taxi services, hotels, and food production. It wasn't until the 1970s and 1980s that Nintendo found its true success by pivoting to video game design and manufacturing, creating iconic consoles like the Nintendo Entertainment System (NES) and popular game franchises like *Super Mario*.

Today (2023), Nintendo's market capitalization is around U.S. $60-70 billion.

Do you see how these three qualities support the others when you pivot? You must *persevere* to support your values *intentionally*, and you *intentionally* may need to learn a new skill, launch a new product, take a course, change direction, occupy a new role in the company, or launch a new company—as you demonstrate *versatility*.

That's why this next point makes so much sense. The person who makes a Pro-Level Pivot does so because he or she is:

Open-Minded

There are two parts to this: When I am open-minded, I welcome new ideas, even if they aren't my own, and I remain humble enough to accept guidance.

This is the part of pivoting when you brainstorm ideas with others. This is when you seek advice. This is when you hire a coach. This is when you allow for the possibility of doing something you've never done before.

> ***This is the step where you imagine the future as it could be, not as it is.***

Don't miss this. This is the power of imagination. We have spent a lot of time discussing this. Before anything that IS ever was, it first existed in imagination, including you.

We must be open-minded enough to think:
What if...
How about this...
Have you ever considered...

Now, admittedly, this is hard. Do you know why? Remember, by age 35, we are living life in the realm of logic and memory. Before that— especially in childhood—we still lived out of active imagination.

We must shift from someone who just remembers or just maintains to someone who imagines. It's a mindset shift. This is why I love the quote, "Never give up on your daydream." That's the stuff imagination is made of.

When we live in memory, we move *from* the past. When we live in imagination, we move *to* the future. Big difference.

Time Management

This all fits together nicely. You already know that you can't do everything. So, as you practice *Perseverance*—forward motion—you do so in an *Intentional* way. You're flexing and diversifying, capitalizing on new opportunities. This is *Versatility,* and it's also *Open Mindedness.*

And as you envision all the possibilities, even new ones, this requires wise use of your *Time.* High-performance people know this. You can't do everything, so you must use time for what counts the most.

An example is found even in how you begin each day. Did you know that the way you spend the first twenty minutes of your day sets the tone for the whole day?

If you hit the *snooze button* for your first twenty minutes, you are dragging yourself reluctantly into the day and not energetically anticipating the potential of what's coming.

If the first thing you do when you open your eyes is grab your phone and read the news or scroll mindless social media, that sets your mind (mindset) on things that would never make your values list. You've already wasted the precious commodity of time.

But if you get up, set your intention, express gratitude to God for another day, begin to consider the alignment of your schedule with your values, and imagine your day unfolding the way you desire, this is alignment. This sets you on a course for success.

> ***There are 1,440 minutes in a day. Just 20 of those can radically change how the entirety of your day goes.***

Pivot!

When you pass through the *death of* the way you *thought* it would be, accept that death, and then envision the way you want your future to be, you move beyond the failure, the place you have been stuck, the place you've been disappointed, the death of your dream. You face the unexpected turn, and you emerge with a life that's so much better.

It took introspection, some time with a therapist, and raw honesty about where I was in life and where I wanted to be before I fully grieved my losses (expectations of how I thought it would be) and stepped into the future that was waiting on me to arrive. It was work, but it was worth it.

> ***Life doesn't have to be the way you thought it would be in order to have a great life.***

In 1928, the Scottish bacteriologist Sir Alexander Fleming was working on experiments involving *Staphylococcus* bacteria, which can cause infections in humans.

During this time, Fleming went on vacation and accidentally left a petri dish of *Staphylococcus* bacteria uncovered on his laboratory bench. When he returned from his break and checked the Petri dishes, he found an unusual mold growth in one of them. This mold, later identified as *Penicillium notatum*, had seemingly killed the surrounding bacteria in the dish.

Although his experiment had been unintentionally contaminated, Fleming had made a groundbreaking discovery: The penicillin mold produced a substance that killed various types of bacteria, later known as an antibiotic.

His experiment turned out *nothing* like what he expected, but this unexpected outcome paved the way for the development of antibiotics, a class of drugs that has saved millions of lives by treating bacterial infections.

Is your life not the way you thought it would be? You are not eternally bound to that place. You are not consigned to suffer in that circumstance. As

I stated earlier, it took time and work on my part to process my losses and climb up to a place where I could see a new future. Allow the old expectations of your life to pass away, and then pivot and face the possibilities for what's in store in your brand-new, unexpected, and exciting future.

CHAPTER ELEVEN

The Death of Who Others Expected Me to Be

If the last chapter was all about dying to disappointment, this chapter is about dying to "people-pleasing." I would ask how many "people-pleasers" are reading these words, but you would raise your hand just to please me.

Let me give you two quick scenarios to set this up:

A teenage boy grew up in a home with his brothers. All of them were extremely gifted athletes except the youngest boy, Joe. Joe was highly artistic. He was an excellent photographer and painter. He listened to music in his room and connected to it emotionally. Any creative expression captured his heart and his attention.

Joe was also on the high school baseball team; not that he wanted to be, but it was an expectation. His older brothers were gifted baseball players, and one of his brothers got a scholarship to play ball in college.

Joe was just average at baseball. And worse, he hated playing the game. The only reason he did was to please his parents.

*****

Mary Jo was the mom of a daughter and son, Irish twins just 15 months apart. She loved being a mom. She engaged with her kids' friends as she drove them all to activities. She attended her kids' events and was a "room mother" in their classrooms.

When her oldest graduated from high school and left for college, Mary Jo adjusted because she still had one child at home, but the next year, when her youngest left for college, Mary Jo went into an emotional funk.

For the better part of two decades, Mary Jo had found her whole identity as a mom, but when she wasn't needed all day, every day in that role, she found herself asking: *Who am I?*

*****

Think of your identity. Is it tied to what someone else expected you to be or hoped you'd be, or even *pressured* you to be? Do you feel as if you're defined by a label like Mom, Husband, Sibling, Athlete, or perhaps the job title you hold?

Wearing a label is like driving down Main Street in any city and seeing new businesses occupying old storefronts when all that's changed is a new sign hanging above the doors.

Throughout your life, multiple signs have been hung on you, too. Some of the signs fit comfortably. Joe loved wearing the sign that said ARTIST, but he chafed under the sign that said ATHLETE.

Mary Jo loved the sign MOTHER, but she had no idea what sign to wear in the next season of her life.

And this is also true for each of us. We have all worn signs people placed on us that didn't fit all that well, but we wore them to please others, to fit in, and at the deepest level, we wore signs we didn't want in order to be loved.

"PEOPLE-PLEASING"

The good parts of "people-pleasing" show up when kids learn to be obedient in school, follow the rules, and respect authority. "People-pleasing" works in the corporate culture where you fit into the culture in order to ascend.

Can we chat honestly? Many of you who are reading this right now will realize that you became a "people-pleaser" just to survive. You had a bully for a parent, or a boss, or a coach, or a teacher, or a brute in your friend group. You became a "people-pleaser" out of necessity, and, for most of you, you didn't even realize it was setting a pattern in your life that would continue for decades.

And though it may have started to survive, it stopped serving you years ago. At a toxic level, "people-pleasing" is now driven by deep-seated fears of rejection, abandonment, or not being good enough.

Here's a definition I wrote that we could work with:

"People-pleasing": Compromising or abandoning who we are, what we want, or what we value in order to be validated, included, or loved.

But oh, what a price you pay. According to a couple of studies, chronic "people-pleasers" experience depression, anxiety, and dissatisfaction.[1]

In the movie *Runaway Bride*, the character played by Julia Roberts dates a few different people in the course of the story. A series of scenes show her ordering eggs at a restaurant with each of these different dates. Each time, she orders her eggs the way her date orders his. The subtext is that she no longer knows how she likes her eggs; she orders in a way intended to please the person she is with each time.

Few things will leave you more exhausted, emotionally unsteady, and falling short of your dreams than "people-pleasing."

---

[1] Hewitt, P.L., Flett, G.L., & Mikail, S.F. (2017). *Perfectionism: A relational approach to conceptualization, assessment, and treatment.* Guilford Publications.

Let's walk through the reasons why. Here are my top ten reasons "people-pleasing" is so damaging:

1. **"People-Pleasing" *Limits* How High You Can Climb**
   Few things are more destructive to your dreams than your addiction to someone else's dreams for you. It's exhausting. You don't have the energy to both please others and achieve your highest. Imagine climbing a mountain with every person you are trying to please on your back. It's both miserable and impossible.

2. **"People-Pleasing" *Abandons* God's design for you.**
   Whatever you believe about your ultimate origin, please allow me to share my faith tradition. Rather than God seeing you as broken, limited, or lacking, we find these words written about you:
   "*I am wonderfully made (by God)*" (Psalm 139:14)
   "*We are God's masterpiece*" (Ephesians 2:10)
   When we become something we are not in order to please someone else, it's like we are denying the divine touch each person possesses. Whether you believe in God or not, most people know we really didn't accidentally come from sludge and then crawl up on land and develop from this *accident* into a being that can create art, construct skyscrapers, develop AI, or stand in awe at sunsets. Something was placed in you that originated from a divine touch.

3. **"People-Pleasing" *Shrinks* Your Dream**
   You only have time for one destiny. Within you is a destiny trying to emerge. If you continue to read off the script your parents, boyfriend, or girlfriend wrote for you, you will play their character. But if you read off the script that comes from your heart, your life, your mind, and your dreams, then your life soars. Someone's dream for you always causes your own dream to shrink if you allow it to.

4. **"People-Pleasing"** *Robs* **Intimacy**

   How can you get close to anyone when all you offer is an avatar of yourself, a fake you, the you that you pretend to be in order to be loved? And here's the built-in contradiction. You "people-please" in order to be loved, and you do it because you're terrified you won't be loved. How can you possibly have true intimacy in that relationship?

5. **"People-Pleasing"** *Diminishes* **Alignment**

   Your true self, your true emotions, your true personality, your true dreams, and your true energy must all align to reach your highest potential, and none of those attributes are present when you "people-please."

6. **"People-Pleasing"** *Avoids* **Conflict Resolution**

   Is there any wonder why "people-pleasers" have anxiety? All of us have had an experience of wanting to say something, wanting to share where we were hurt or disappointed or lied to, or (pick a time you didn't speak up in order to please someone else or avoid displeasing them). Yet, to avoid conflict and maintain our fake persona, we bury the emotions. They live within us and come out in negative, unintended ways.

7. **People-Pleasing** *Destroys* **Confidence**

   "People-pleasing" arises from fear. Fear is the dominant emotion "people-pleasers" feel. And confidence gets smothered by fear.

8. **"People-Pleasing"** *Increases* **Stress**

   "Does she like me?"
   "Did I do okay?"
   "Did he smile?"

"Does she approve?"

It's exhausting and riddled with stress!

9. **"People-Pleasing"** *Chases* **the Impossible**

    The truth is you won't please everyone. But you will kill yourself trying.

10. **"People-Pleasing"** *Fails* **Jesus' Command to Love**

    Again, you likely know this one, even if you are not religious. Jesus issued this very famous statement: "Love your neighbor as yourself." How can you love your neighbor as yourself if you haven't yet figured out how to love yourself?

    Or, as I like to say, if most people love their neighbor as they love themselves, their neighbor is in deep doo-doo.

Here's the fundamental question: How do we die to the "people-pleaser" in us so we can become our authentic selves?

Here's the first step. Any change in your life is successful to the degree you believe it's possible. There's the iteration of you, which is who you currently are. Do you believe you can re-iterate?

It's like the Apple iPhone. As of this writing, Apple has had thirty iterations of the iPhone, with more to come.

There should at least be that many of you. You aren't the same person as you were when you were five, or fifteen or twenty-five. And you can have that much change or more in the coming months. What is stopping you from becoming YOU version 2.0, 3.0, or 30.0?

If you're tired of pretending and want to finally emerge as the wonderful you that you were made to be, change! Let me give you a mantra to repeat every day to bolster your confidence and belief:

If it isn't working for me, I can change!

There was a department store named "Dayton's," then it was "Dayton-Hudson," until the company underwent a major rebranding effort in the 2000s to appeal to a broader audience. The rebranding included the introduction of a bulls-eye logo, which has since become the highly recognizable symbol of the Target brand. If a department store can change, so can you!

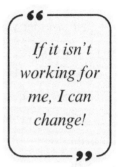

If it isn't working for me, I can change!

Change!

Before Dwayne Johnson became one of Hollywood's biggest stars, he was a professional wrestler in the WWE. His success as a wrestler was largely driven by the persona he created: a tough, no-nonsense fighter known as "The Rock."

However, after several years in the wrestling world, Johnson realized that he was no longer passionate about the sport and that he was primarily doing it to please his fans.

Despite the initial skepticism of Hollywood executives, he left wrestling behind and pursued what he really wanted to do: acting. Since the change, Johnson has become one of Hollywood's highest-paid actors.

You may be asking, "But Brad, how can I change when I want approval—or deeper, how can I change when I base my value on validation from others?" This is why we must keep going on this topic.

But believing you can change is only the first step. Add to that your belief in your innate value as someone created by God and born with greatness already in you.

I only share from my faith perspective in hopes of encouraging you. For instance, did you know the ancient Hebrew scriptures teach that we are each created in God's image? The divine imprint and shape are upon you and within you. That, all by itself, gives you infinite worth and value.

This means you have worth as the object of God's love. The story of God is the story of a Father who loves His kids.

No wonder you have the feeling you were made for more. Do you want to know why?

Because YOU. WERE. MADE. FOR. MORE!

Here's the next part: Look for every congruent "Yes!"

When you begin to live authentically, you will make your best contributions with the opportunities that fit who you authentically are. So, every time you learn what you should say *yes* to, you have also helped yourself by defining what you will say *no* to. You will recognize where you should draw a line and determine that there will be no more appeasement and no more pretending.

Saying "Yes" to what is congruent to your true self and "No" to what isn't your best fit is called "Establishing Boundaries."

Maybe you will end up saying "No" to a job offer. Maybe as an adult, you will say "No" to a parent or to a friend group. It's easier than you imagine, and you will find tremendous empowerment when you can honestly say: "No, that's just not me."

When you do that, you give yourself the gift of space, and capacity, and freedom to offer a passionate "Yes" to the people and opportunities that align with who you are. This is all about standing up for yourself, expressing yourself, and expressing what you need and want.

Here's an example of a congruent *yes*. Karen and I are not night people. We just aren't. We aren't "go to L.A. and do something late at night and get home at midnight" kind of people. Our whole friend group knows that if an event or activity is going to keep me up past 8:30, I am going to say, "NO."

Allow me to explain further. My burst of creativity is in the mornings, beginning about 4 a.m. The energy and electrical frequencies of my brain and heart are humming at those hours. I get so much written, so much created, so much planned between 4 a.m. and 8 a.m.

So by 8 p.m., I'm thinking seriously about getting horizontal for the night, and by 8:30 p.m. I'm usually in bed reading or keying down.

I've wanted to be a late person because most people are. I've wanted to be a "go out and have fun with friends or a late dinner or concert in LA kind of person" because most people love that. But every time I am untrue to my own biorhythm, every time I pretend I'm a night owl, it's not just the next morning that suffers; it's a couple of days of less creativity, less contribution to the lives of people I want to help, and less of me.

Sometimes, Karen and I get lovingly teased about this. But guess what? I'm so much happier when I am true to who I really am. Does this make sense? I love *me* more than I love a late night in the city!

Once you feel you have confidence in living these first steps we just covered together, begin to surround yourself with authentic love. I'm not talking about the pretend love and the pretend relationships you have by compromising your true self to fit in or be accepted. That's bogus. I'm talking about having a few people—you don't need a lot—who know the real you and love the real you!

So much of what we call love is really just shallow transactional business.

We say, "I love you," but what we mean is: I love that you accommodate me.

"I love you," but what I mean is: I love what I get from you.

"I love you," but what I mean is: I need you to fit into the mold of who I need you to be.

You won't find strength or encouragement in that. You will only be your *best* self when you surround yourself with people who love your *authentic* self.

Next, practice habits of self-care.

If you need someone's love more than you love yourself, that's toxic. Something happens in your spirit when you aren't who you were created to be because that's not loving yourself.

Set your values and live by them. Set your intention—like we talked about in the previous chapter. Believe you were created for more and given everything you need for your destiny. Set boundaries like we talked about (saying YES and NO). Spend time with your lovable self. Check in with yourself to see how you're doing—to assess what you need and to stay clearheaded on your purpose and dreams. Make time to exercise, eat healthy, spend time with loved ones, invest in your faith, get enough rest, and engage in activities that bring you joy.

That means:

Show up for yourself!

Some of you right now are exhausted because you show up for everyone else except you. You know you are made for more, and you desire more, but you're so darn tired from "people-pleasing" all day long. You must understand that no one else is coming to show up for you if you don't lead the way. Show up for yourself.

I know this feels scary. You have years of *not* showing up for you. It sounds daunting to do so. Your heart is racing, and you likely want to know what will happen if you do.

First, your mental health will improve significantly! People who live less than their authentic selves feel depressed, anxious, and stressed because they are living out of alignment with who they truly are.

Living authentically allows us to be comfortable with ourselves and at ease in our own skin. This leads to increased self-esteem and an overall sense of well-being.

Do you want to know what it feels like? Peace! How long has it been since you had a contented feeling about yourself and about your life? Here's great news: It's within you to experience that. Your path to peace is found when you show up for you.

You will also experience deeper relationships. Why? Because, after living so long in fear that if someone really knew you, they'd reject you, something fantastic will happen when you learn to love yourself and show up for yourself.

When you find your voice and you find your way, someone is going to love the *you* that they see. And then another will love you just the way you are. Then another.

And it won't be fear-based: *Will they or won't they like me?*

And it won't be stressful: *If I don't keep up the charade, all this will go away.*

It will arise from a place of loving yourself. You send out confidence, an energy that signals to everyone in your orbit that only those who love you just the way you are are welcome. And that is who will start showing up.

There is another part to the wonderful changes that come when the "people-pleasing" part of you passes through a death. You will achieve greatly and fulfill your destiny.

It takes full alignment to become your best self and fulfill your destiny. All of you must be pointed in the same direction—the direction of the life you've dreamed.

All of you must align: all of your dreams, all of your beliefs, all your emotions, all of your values, and all of your actions. When they do, there's great power surging from you and propelling you into your future.

And people who don't get it or don't get you, who maybe even say, "You're just too much," won't get to go with you to your new, soaring heights. And you will be okay with that.

If you've tracked Lady Gaga's career, it appears meteoric, but it wasn't always. She was pushed and pressured to not be theatrical, not be so avant-garde. She has talked about how she had to overcome her "people-pleasing" tendencies to find success in the music industry. She learned to prioritize *her* artistic vision instead of trying to please others, even when it seemed risky or unpopular.

You must combine everything that it means to be You, align everything in you, and point your life toward your goals and destiny. Then, you will find the power to achieve greatly.

This week, someone who didn't even know I was working on this chapter wrote to me. She asked how she could tell if she was serving other people in her life from an honest and authentic place or from a place of "people-pleasing."

Here's what I wrote: *"Great question. It all comes down to the WHY and the WHO."*

I cited Jesus as an example (of course I did): *"Jesus served from a place of love for others. The WHY wasn't to earn acceptance or appease or please. It was purely to love.*

"And the WHO involves knowing WHO you are. Or are you playing a role that isn't really you? Authenticity is the goal. From a pure heart of love, bring your authentic self to others, and that is how you know if you are serving from a good place or from the toxic base of people-pleasing."

Bring your best self forward if you want to have influence and serve others. Bring your authentic self forward to have the greatest impact on your one and only life.

Let me give you an image of what your best life looks like. It looks like a bright light shining from within you into the lives of people going through a hard place and a dark season. When you are fully who you were created to be, you bring light to the darkness, hope to the hurting, and encouragement to the discouraged. See, too many people are living by default and not by design.

I must confess something. For too long, I hid my light, or at least diminished it. I lived by default, not by design.

Because of my past, I didn't feel fully worthy of shining my light, so I kept my head down and stayed in my lane, as I described earlier in the book. And further, I feared the criticism that comes whenever you begin to show up as your authentic self.

I kept my light under a basket; I feared stepping out, and I just hoped that my diminished light would please people. But that stirring that some of you feel right now, that stirring of being made for more, grew and grew, challenging Karen and me, convincing us that we had bright lights the world needed.

Here's the full truth: The world needs my fully authentic, full light! And the world needs your fully authentic, full light.

> *The world needs my fully authentic, full light!*

Say this out loud wherever you are right now:

The world needs my fully authentic, full light!

That deep-seated conviction set Karen and me on a journey, wondering how to expand our influence, encourage and lift more and more people, and shine our light brighter than we ever dared before.

Parts of what all that means are still being discovered and created and revealed in my life, but here's what I know for sure: My best contribution to others is when I show up *for* myself *as* myself, fully authentic, shining my *full light*. And that's absolutely the truth about your life.

I am going to ask you a question, and it's the most important question you've been asked this year, maybe ever: What happens if you live every day the way you are living right now, and when you die, you talk to God? Suppose God slides a piece of paper across the table. You read it and see the description of an incredibly lived life, and God says: "This is who YOU were supposed to be. This is what YOU could have done with your life."

And then imagine that God asks: "What happened?"

I don't want to have that moment and feel the regret I would feel from not being all I was destined to be, not accomplishing all I was supposed to accomplish, and not impacting the lives I was supposed to serve and influence toward their greatness.

And so, I decided I'm showing up.

I. AM. SHOWING. UP!

And I'm inviting you to show up—not the "people-pleaser" you, not the public fake persona you. Let your full light shine. Then, let's go out and change the world.

CHAPTER TWELVE

The Death of Fear

Why am I taking so much page space to talk about these deaths? Because each of these four chapters, tacked right into the heart of this book, identifies specific areas that can steal your dreams, keep you from your goals, and hold you back from your destiny. And conversely, navigating through these deaths has the power to propel your *Bounce Better!* I share these because I care about you. It's important that you learn to pass through these four deaths.

I realize that not everyone wants that; not everyone aspires for more. Not everyone has a flicker of fire burning on the inside, waiting to be released to serve the world in a greater way.

> *Everyone has his or her own journey!*

Everyone has his or her own journey!

That's okay. This is a judgment-free zone.

But for those who have lived with this uncomfortable feeling that you are made for more, you want to achieve and accomplish and grow into everything your creator put you on Earth to experience and achieve, keep reading.

You have already been given the tools to pass through the death of disappointment—when life didn't turn out the way you thought it would, when you find yourself far short of your hopes and dreams.

You learned in the last chapter how to pass through the death of "people-pleasing" when others try to put you in their box and you allow yourself to read off the script someone else has for you.

In this chapter, we jump into something that is as common as breathing. It influences every person reading this—sometimes way more than it should. And talk about keeping us small, limiting our future, diminishing our influence and impact on the world—this one is huge. It's FEAR. Through the years, I've asked thousands of people this question:

If you could do something with your life and you knew you wouldn't fail, what would you do?

The very thing you just named is likely what you *should* do. It is the *something* that is burning deep inside you, a hope and dream, an ambition, or even a deep sense of calling in your life. It's that unshakable, unforgettable thing, that If-I-could-I-would thing.

Imagine yourself right now doing it without fear. Imagine yourself right now doing it with confidence and boldness and effectiveness. Imagine yourself achieving it.

Why don't most people think this way or *live this way*? It's because most people settle, and here's why most people settle: life beats them down and beats them up. They tried something once, and it didn't work out, so they didn't try again. They wanted to do something bigger with their life and told somebody about it, and that person made fun of them or shamed them, and they gave up the dream. People settle because they stop believing in the dream and/or they stop believing it can happen for them.

Life has a way of shrinking your dream to match your faith. Let's be honest: Life has a way of beating confidence and courage out of you.

And most of the punches thrown at you are thrown by the enemy of fear.

I must write about fear because most believe it's normal, even good. They say things like, "Brad, isn't fear normal?" and "Doesn't fear protect me from harm?"

No.

Ordinary caution is normal and protects you from harm. Common sense is normal and protects you from harm.

But fear limits. It's fear that keeps you from even trying.

Fear diminishes who you are and what you can achieve.

Then I meet people who live afraid, and they justify their fear with what sounds like a super spiritual question. They ask, "But doesn't the scripture say, *'The fear of God is the beginning of knowledge'*? (Proverbs 1:7)"

Yes, but the word translated as "fear" is more correctly translated as "respect."

Let me offer an example of the difference. I see this in scuba diving all the time. People who fear the ocean often don't continue getting in. Or they are so tentative as to be a danger to themselves or others.

However, those who respect the ocean approach it with an understanding that it's

powerful and life-giving, and it exacts a high price on those who are careless. Those who respect the ocean carry that respect and still step into the water. There, they experience all the depths and wonder of the ocean and all its magic, and they immerse themselves in it.

Now, replace the word "ocean" with the word "God." Those who fear God—in our understanding of the word fear—stay away from God, they won't approach God, they cower and primarily have negative associations about God.

But those who respect God, understand His power, know He's life-giving, and are not careless or disrespectful of Him draw close to Him and experience all the depths and wonder of Him.

Respect allows us to explore and go deeper, but fear keeps us on the shore and far from what awaits us in the water.

People have all kinds of fears. According to the National Institute of Mental Health, specific phobias affect an estimated 19.2 million adults in the United States.

In a 2018 study by the American Psychological Association, more than one-third of adults in the United States reported feeling afraid or anxious about personal safety or the safety of their loved ones.

With the recent pandemic, people fear for their own health and the health of their loved ones.

Financial instability, job loss, and economic recession are other major fears among Americans. Many people fear they may not be able to support themselves or their families financially.

A huge category of fear is the fear of judgment. Here's an example. Most people reading this can relate to the fear of public speaking. Do you know what that really is? It's the fear of judgment. We don't fear that words won't come out of our mouths. We fear that the words we speak will be judged; our ability to clearly articulate what we want to communicate will be judged.

And this fear of public speaking isn't just the fear of speaking from a stage. It includes expressing yourself in a crowded room, asking for what you want in a relationship, or offering your opinion in a group discussion. But let's dig even deeper.

Behind the fear of judgment is the fear that we are not enough!

"Not-enough-ness" has killed more dreams and stopped more dreamers than failure ever will. And this "not-enough-ness" shows up all the time.

With public speaking, it's the fear that what we have to say isn't enough.

As parents, we fear we are not enough to do it right.

As we face retirement, we fear we aren't financially ready or enough.

As we seek a better life, we fear we can't because we are not enough.

As we move out of our parent's home, we fear we are not enough for independence.

As we seek to succeed at work, we fear we are not enough.

As we desire to start a new relationship, we fear we aren't enough to be chosen or loved.

More and more, I have such clarity on this as it relates to the rest of my life, and this is a fundamental purpose for this book you are reading right now:

My mission is to help you know you are fully enough, to help you see your great worth and to coach you to achieve your God-given potential.

My highest joy is to see someone embrace their *enough-ness* and step into their true destiny. Here's what I am absolutely convinced about: You were born for success, not failure and not limitation, but over time you have been conditioned to settle.

I am committed to changing that. I want to infuse belief into you. And why does that matter? Your success will rise to the level of your belief.

Here is a truth that can easily change your whole life:

Everything you need is already in you.

Let's unpack this together.

God's Spirit is in you.

We already learned that each of us is created with the imprint of God's image within us. This creative, powerful, loving God made you, and He didn't make you for no reason. He didn't make you to accomplish nothing. He didn't make you to have no impact in service to others. He made you for a mission, and that will be discussed further in the next chapter.

> God's Spirit is in you.

I don't know what you believe or understand about God, and I do not want to impose my belief system on you, but I need to show you something that is said in the scriptures about how God made you. This may be the most important thing you'll learn today:

> *"For God has not given us a spirit of fear and timidity, but of power, love and self-discipline."*
> (2 Timothy 1:7)

That means when you feel fear, it doesn't come from God.

What else is in you? The dream you carry and the feeling that you are made for more.

What you desire comes from your own background, passions, personality, gifts, talents, and experiences. That all shapes what you want from life. The fact that you have dreams for more should be enough to convince you that you are made for more.

YOUR BACKGROUND AND EXPERIENCES

No one has had the exact same life experiences as you. No one has the same perspectives, the same wisdom, the same insight, the same history, the same relationships you have lived through. And you carry all that combined as a valuable asset for your future.

Your education earned the hard way. Have you learned more through the easy experiences or hard experiences of life? Hard things help us learn in ways that ease and comfort cannot.

Your imagination. I simply cannot overstate the importance of imagination in your life. It singularly influences your future. The moment you stop imagining your future as you want it to be is the moment fear wins. It's the moment your past wins. It's the moment you've stepped into cement boots, and you're stuck.

If you feel stuck and want to get moving, start imagining the life you want to live.

Your intuition. Some people think this is woo-woo, but listen, intuition is a combination of non-verbal cues you pick up on in any given situation. You are analyzing in your subconscious mind everything you sense and see, and your mind interprets that information. Think of intuition as super-computing. Your mind is assimilating information in every circumstance of life.

Intuition is part experience, part perception, part wisdom, and part spiritual. We must learn to listen to our gut way more than we do.

Your hunger. Only you know how badly you want more. Only you know the fire burning in your spirit, the desire in your heart. Your hunger is your ally; it's your fuel.

Your values. Earlier, you learned about Values Based Intentions. This is your intention to serve what you say matters in your life. This is your WHY. This is why you do what you do. This is what truly holds value for you. Values get you up in the morning and help you push through fear.

Your attitude. Seeing the future through a positive lens or a negative lens is completely your choice. You can choose your attitude. Certainly, you can't choose all your circumstances, but you can choose your attitude. This attitudinal choice is within you.

Your free will. You get to choose what you pursue. You are not a slave to the past or to your circumstances. If you want to change from where you are to where you want to be—*MOVE!* You are not a tree. You can re-imagine and then freely choose to pursue the future you will live.

Your creative power. The ability to bring something from nothing is what sets us apart from the animal kingdom. We can imagine, and then we have the power to create the future.

Karen and I have the sweetest, smartest dog. But little Pepper can't imagine and create. He can't look at something he enjoys and understand that what he sees, or plays with, or even eats once existed only in imagination. But this creative power is already in you.

You see something in your mind, and you can build it.

You see something in your mind, and you can begin to assemble a team to bring it to reality.

You see something in your mind, and you can write it down. That immediately makes it tangible. You move it from thought to paper, from paper to action steps, from action steps to work, and from work to manifestation. That is the power of creativity, and it's already in you.

Your mindset. This is what my high school wrestling coach called mental toughness. It's getting your head right. It's removing the inner critic, silencing the inner voices that don't believe in you, and replacing those voices with a constant stream of faith—a constant stream of *can do*.

Every person reading this book has every one of these powerful attributes at their disposal, already inside. But do you know what can negate each one? Fear! Fear that you are not enough.

Over time, if you allow fear to accumulate and be a constant presence, your continued fear will break down your body. Cortisol levels rise, adrenal fatigue sets in, and, worse, fear breaks your spirit.

There is ongoing research into the relationship between emotions, including fear, and the measurable electronic frequencies in the mind, heart, and outside the body. You can literally feel your fear outside your body, and others can feel it, too. Your body expresses the fear you carry. Your body is keeping score.

When it comes to leaning into your destiny, even trying to find your destiny, even exploring your purpose, or defining your dreams, these are all thwarted when you live afraid.

How much good is never attempted because of fear?

How many life-changing conversations are never voiced because of fear?

How many career-catapulting actions are never taken because of fear?

Let this truth settle into your heart:

Everything you ever wanted is on the other side of fear.

That healthy relationship you didn't attempt because you were afraid to show up; that career path that makes your heart soar and you've been so afraid of embarrassment or failure to try; the secret dreams you have been so afraid to voice, let alone pursue—These all exist for you just on the other side of fear.

THE DEATH OF FEAR

I have some terrific news for you. There is a way past fear—putting fear to death in your life—so you can reach your highest potential, so you can *bounce* higher.

Move fear from an emotional response to a logical consideration. When we move our fears out of the realm of emotion, we disarm their strength.

Why? Whenever something is either emotional or logical, emotion wins. It takes great discipline to move thoughts from emotion to logic. Emotion has its place, but not in the realm of fear.

When someone is emotional and a logical person says, "Calm down," is logic or emotion going to win? You know the answer, and so do I.

As you think about the dreams in your life, the career you really want to pursue, the relationship you really want to have, the conversations you really wish you could have, the business you really wish you could start, the money you really wish you could earn, you disarm your fear when you start asking questions like:

- Why does this frighten me?
- What would the most successful person I know do in this situation?
- What are two steps I could take this week toward the life I truly want to live?

When we break fear down in a way where we question and analyze, our heart rate decreases, our breathing evens out, and the power of our thoughts—

our imagination—begins to propel us toward solutions and actions instead of spinning in emotion.

Lean into the power of prayer. I know not everyone reading this is a person of faith, but millions of people have found strength and courage through prayer. Those in recovery programs acknowledge a higher power. Many scientists and scholars, athletes and celebrities, mothers and fathers, the elderly, and the young all have experiences of comfort, guidance, and power that came from prayer.

When we get to the end of ourselves, when we've worked as hard as we can, when we've thought as much as we can, we will still need a higher power.

Christianity is central to everything for me. So, I begin my day praying from that perspective, mainly in the mornings, and I start with gratitude. This moves me from a place of scarcity, which is fear-based, to abundance, which believes there is plenty.

Several studies have examined the role of prayer and meditation in facing or overcoming fear. A 2019 study published in the *Journal of Religion and Health* found that meditation and prayer can significantly reduce anxiety symptoms in people.[2]

Trusting in a higher power can help people overcome their fear.

Here is a metaphor from an experience I had recently with a frightened scuba diving student. We were in a pool that started shallow and slowly descended to 12 feet. I started with the student in the shallow end. We had all our gear on, and we submerged in 4 feet of water. The plan was to swim along the bottom from the shallow end to the deep end and swim around there. Once we submerged, she took my hand and wouldn't let go. And guess what? She stayed for the whole class, swam underwater for thirty minutes, and accomplished her goal!

---

[2] Winkeljohn-Black, S., & Meisenhelder, J. B. (2019). Unpacking the Relationship Between Prayer and Anxiety: A Consideration of Prayer Types and Expectations in the United States. Journal of Religion and Health, 58(1), 356-370.

Why? Because she held the hand of someone who stayed beside her, someone whom she trusted to help her. When we pray, we are reaching for a power outside of ourselves, toward one who will stay beside us and help.

Keep calm and take action. Let me give you one more example from scuba. The number one rule of scuba is—keep breathing.

Let's suppose something happens under the water that causes a moment of fear. The first thing you do isn't shoot to the top of the water or frantically wrap yourself around another diver. The first thing isn't even to solve the problem. The first thing a frightened diver is taught—keep breathing!

Keep calm, then take action.

One time I was diving off Catalina Island at about 40 feet under with bad visibility, and my diving buddy and I lost sight of each other. I couldn't see him in the murky deep and he couldn't see that I got tangled in kelp. What's the number one rule in scuba? Keep breathing.

We breathe, then we act, we breathe, and then we act.

In a diving situation, focusing on our breathing calms us, our mind slows, then we can make good decisions. So, with careful breathing, I was able to take off my gear. I untangled myself and soon was enjoying the rest of my dive.

As you assess the fears that hold you back, that keep you playing small with your one and only life, that hinder your relationships from going deeper, first breathe—keep calm.

Here is an exercise to try right now, and then spend time over the next couple of weeks perfecting this in your life. Get in your mind what is stressing you right now—a relationship that's broken, a business plan that's faltering, financial pressure that's crushing. Now, I want you to practice "mental breathing."

> *Mental breathing is visualizing a positive outcome for the area where you are currently afraid.*

Visualize a positive outcome. Some people teach you to consider: *What's the worst that can happen?* Here's my experience: People are already doing that, and that is why they are terrified. Instead, begin to imagine: *What is the BEST that can happen?*

Are you afraid of trying to lose weight? What does it look like if you do reach your goal weight?

Are you afraid to try a business venture to work toward financial freedom? What does it look like if you do earn your dream income?

Are you hoping for a loving relationship? Instead of *I'm afraid no one will want me,* imagine instead that you do create a beautiful relationship. What would that look like? What would that feel like?

Let this better way of thinking flow in and out of you like life-giving oxygen. Breathe in that positive, hope-filled image over and over. This is mental breathing.

Exhale the fear. Inhale faith. Right now, I want you to take sixty seconds and, with the following prompts, consider your best life. As you do this, release fear over your future and inhale these images over and over.

Imagine yourself:
- Growing Stronger
- Living Fully
- Loving Deeply
- Succeeding Amazingly
- Impacting More Lives
- Helping More People
- Showing Up for Yourself
- Showing Up for Your Family

Mental Breathing empowers you to do two things: 1) Define your *intention.* This is a purposeful aim, and 2) Begin *acceleration.* This is an action of movement.

To get past fear and begin moving in the direction you desire for your life, there must be a target to aim at, an intention. And there must be action—acceleration.

"Brad, you mean take a step while I'm still afraid?"

Yes! Because *an unsure step forward is still a step forward.* If we don't take aim and act through intention and acceleration, here's the next part of a predictable process:

Stagnation: That is a lack of progress

We have seen this in sports. A team is playing with bold confidence, driving, and scoring. They are winning in every measurable way. Then, their mindset shifts from playing to win and to playing not to lose. When you play to win, you still see a great future, you work for more points, a stronger lead over the opponent, and an exciting outcome. But when you play to not lose, you actually feel the momentum die… and the main goal is running out the clock.

> *The main goal of playing not to lose is running out the clock.*

The main goal of playing not to lose is running out the clock.

If someone were to measure the energy of your life, the excitement of your life, the gleam in your eye, would they find you playing to win, or tragically, would they find you just running out the clock? If you're running out the clock, it's not too late to get back on offense and play to win.

Here's the pattern: Intention, Acceleration (playing to win), Stagnation (play to not lose), then comes *deterioration:* a slow, certain death

Do you want a life of meaning and purpose, of impact and significance? Do you deeply believe you were made for more, and you want it more than anything? It's all within you to have, but remember:

Everything you ever wanted is on the other side of fear!

Malala Yousafzai is an inspirational figure who faced fear and achieved greatness. She was born in Pakistan in 1997 and became an activist for girls' education at a very young age.

When Malala was just 11 years old, she started writing a blog under a pseudonym for the BBC about her life under Taliban rule in the Swat Valley. She was an advocate of education and wanted all girls to be educated, which was not allowed under Taliban rule.

In 2012, Malala was only 15 years old and was targeted by the Taliban, who boarded her school bus and shot her in the head for speaking out against their actions.

She was airlifted to the UK for treatment. Beyond anyone's hope, she miraculously recovered.

Despite the assassination attempt, Malala didn't stop fighting for what she believed in. She continued her education and became an *even more outspoken* advocate for girls' education. She started the Malala Fund, which aims to help girls get access to education in developing countries.

In 2014, she became the youngest-ever Nobel Peace Prize laureate. And she was just a *child*, but that young lady played to win!

"God will not have His work manifest by cowards."
–Ralph Waldo Emerson

Can you imagine your life on the other side of fear? Can you imagine your life on offense, playing to win? Stop dimming your light because you are afraid. Stop silencing your dream due to fear. Stop worrying about what could go wrong, and lean into a faith that imagines what if it all goes right!

Your best life is waiting, and it's just on the other side of your fear.

CHAPTER THIRTEEN

The Death of Living Small

This chapter concludes with a mini section on the Four Deaths that it's necessary to pass through. What an ironic subject for the very principles that will lead you to your greatest life.

I offer these chapters so that you can review them every time you are feeling stuck, confused, or stumped on how to get started again. I've seen these same principles take me from being 17 days away from homelessness, completely broke, with no food in the refrigerator, shattered in spirit to the point of suicide, all the way to watching as my life bounced back.

But, to get to the life I am living today, I had to pass through each of these four deaths. This chapter is vital: Death to Living Small.

Let me tip my hand right here. Small living is living just for yourself. That's a small target, that's a pretty small goal, that's a pretty small circle of influence or even significance. Yet, how many people have made their lives just about themselves and just for themselves?

For a couple of years, I did that, and it nearly killed me.

In Christian circles, one of the most successful books in the last twenty years was called *The Purpose Driven Life*. Written by Rick Warren, it sold tens of millions of copies. The compelling first sentence of the book sets the stage for the rest of that treatise on finding meaning in your life: "It's not about you!"

Boom! Your momma may have told you it's about you. Your girlfriend or boyfriend may have told you life is about you. *You* may have told you it's all about you, but the truth is our best life is found when we discover a meaningful mission that serves, lifts, and loves as many people as possible!

Let me set this up with a couple of stories.

Jack just hit 30 years old. He'd finished his MBA and was in full stride toward his goal of earning half a million a year as a stockbroker.

The hours were insane, but Jack believed it would be worth it. Nothing pleased him more than to snag a new, high-net-worth client. Two years later, Jack was rich, but Jack was also divorced, drank too much, and sat in his finely appointed penthouse apartment wondering why the very thing he'd lived his whole life for had left him empty and alone.

He felt like he didn't even know himself, which meant—wait for it—*he didn't know Jack!*

*****

Sylvia had so much potential. She was an Ivy League grad, near the top of her class, recruited to the finest firms in her area of expertise, but Sylvia had another plan.

She joined a Fortune 1000 company but not in the C-suite. She became Director of Corporate Social Responsibility, leading this profitable firm to return some of the profits back into the community, helping underserved neighborhoods in areas of healthcare and education. She had friends, fulfillment, and a perpetual smile on her face, every day knowing she'd discovered the best place for her intellect and contribution to others.

And here's the back story: Her mom never graduated from college, never had a high-impact career, never traveled much, or lived what anyone would call a large life, but when asked about her own meaningful mission, she said, "It's not something I did, it's someone I raised."

BOOM!

Three chapters ago, you learned the Pro-Level Pivot, and here we are with a pivot at another level, pivoting from self-*ish*-ness to self-*less*-ness.

Here's the irony: In helping others, we help ourselves.

I remember working alongside a group of men and women who went to a homeless shelter in the San Fernando Valley, here in Southern California, to cook breakfast and serve a couple hundred men, women and kids who were living on the streets.

A few people in the group had never done this. Afterward, to a person, they each came up to me and said, "I thought I was coming here to bless them, but I am the one who got the blessing." This isn't just a universal, spiritual law; it's backed up by science.

A study published in the *Journal of Happiness Studies* found that people who engage in prosocial behavior, such as volunteering or helping others, reported higher levels of life satisfaction and positive emotions.[3]

Another study published in the *International Journal of Stress Management* found that participants who volunteered reported lower levels of stress and anxiety than those who did not volunteer.[4]

A study published in the *Journal of Health Psychology* found that older adults who volunteered had better physical health and lower mortality rates than those who did not volunteer.[5]

This stuff could literally save your life.

As you assess your life, the direction of your life, not only where you've been, but so much more importantly, where you are going, does a meaningful mission have any space on your radar? Are you asking the important questions, like *What's the meaning of my life? There must be more, right? What does God want from me?*

---

[3] Aknin, L. B., Dunn, E. W. Whillians, A. V., Grant, A. M., & Norton, M. I. (2019). Happiness and Prosocial Behavior: An Evaluation of Evidence. Journal of Happiness Studies, 20(3), 1173-1191.
[4] Hosseini, S. H. & Oremus, M. (2019). The relationship between volunteering and mental health outcomes: Lower stress and anxiety among volunteers. International Journal of Stress Management, 26(3), 259-271.
[5] Li, Y., & Ferraro, K. F. (2019). Volunteering in later life and trajectories of physical health. Journal of Health Psychology, 24(2), 209-219.

When you find your meaningful mission—and I'm going to give you mine in a few minutes—everything changes. Life's colors are more vibrant, your days are more fulfilling, your heart is more joyful, your future is brighter, your moments are momentous, your significance is richer, and you are stepping into your greatness.

That's what I want for you as you explore this chapter. Later in the chapter, you are going to have a working session to help you explore and put some flesh on the bones of any ideas you might have around this for your own life mission.

MEANINGFUL MISSIONS

But let me give you some parameters. Here are aspects that just show up in the great missions of people who are truly making a difference in their lives.

A meaningful mission offers support. Imagine giving a helping hand. Imagine offering a "lift up." Imagine sharing the power of encouragement with someone who is down.

Often, we associate the word support with financial sharing. But maybe the support isn't financial. Maybe instead of a handout, you offer a hand-up. Maybe it's an act of kindness or support with compassion.

A schoolteacher spoke about a boy in her class, and she said, "Every morning he comes in with messy hair, and every day I comb it for him at the beginning of each day before the other kids show up. One morning, when I finished, he gave me the biggest hug and said, 'Thanks. My mom is sick and can't brush my hair anymore, so I'm happy you do it.'"

A meaningful mission offers empathy. I asked ChatGPT for three definitions to explain empathy. Here's what it offered:

1. Empathy is when you can step into someone else's shoes and feel the pebbles they are walking on.
2. Empathy is like a warm quilt on a chilly day—it wraps around you and helps you understand the shivers others are feeling.

3. Empathy is a heartfelt melody that plays in your soul, resonating with the emotions of those around you, and harmonizing your understanding with their experiences.

I liked what one guy said, "But if I walk a mile in your shoes, I'll be a mile away, and I'll have your shoes."

Not quite the point.

Karen and I had the opportunity to go through the Holocaust Museum that came to the Reagan Library in Southern California. We'd visited the one in Jerusalem and we had the same heart-crushing but so important set of emotions going through this one.

There was a quote on a poster from a survivor who said this:

"You had to have a partner—a partner to take care of you—and you to take care of him. A partner who you could organize food with and would share food with. You had to have someone to help because if you stood by yourself, you couldn't survive. You didn't survive. There were those people who didn't want to share anything. They were always afraid to share a piece of bread. And those are the people who went first."

–Mike Vogel, Auschwitz Survivor

The ancient Hebrew scriptures offer this wisdom about the transformation that must take place. God says: *"I will remove from them a heart of stone and give them a heart of flesh"* (Ezekiel 36:26).

This is a beautiful metaphor for the transformation that takes place when one becomes empathetic. Within your chest beats a heart that feels and pulses and bleeds. It's a heart of flesh, not stone.

A meaningful mission shows respect. This is when you begin to see the value in every person, the inherent worth of every man, woman, and child.

I love a concept of humility that I've been toying with that supports this notion. In my definition, humility is not seeing less value in yourself. Instead, it is seeing the value in others. What we've mistaken for humility—remaining quiet, broken, head bent and down—isn't humility at all. The opposite is true.

When we value others, we find an environment for us to show up with our own full value. And just like another's full value blesses me, if I show up small, I have decided not to bless them at all. But when we respect others, we can show up fully with self-respect and make our best contribution to them.

As a child, I was taught to value people over possessions and human worth over net worth. This led me to this truism: If you use people because you love making money, you may get rich, but you won't be fulfilled. But if you use money to help people you love, you may get rich and you will be thoroughly fulfilled.

Millennia ago, King Solomon gave this instruction to his sons: *"A generous person will prosper; whoever refreshes others will be refreshed"* (Proverbs 11:25).

Refreshing another person is born from respect for the other person.

A meaningful mission includes volunteerism. This is the part of lifting and loving and serving others that you do for free. This is the part of your heart that extends with no expectation of return.

Yet, this is something I hear all the time: "Brad, I have nothing to give" or "I'm in a place in my life when I need to receive."

Two things: There's nothing wrong with needing help or receiving help. I have received help many times in my life. But the deeper reality is that's the low-hanging fruit, that's the bottom of the fulfillment ladder. That's not where your greatest blessings or meaning will be found. It's when you give of yourself with no expectation of return that you achieve this.

At my lowest point, I was pretty zeroed in on me. Stuck in my sadness. Stuck in my self-pity, stuck in my cycle of doom and gloom. Then I came across this truth that shifted my perspective immediately: When you feel most helpless, become the most helpful!

There's a dynamic shift that occurs, a seismic change of your energy and frequency and spirit when you stand up to help, when you step out to help. And suddenly, you are no longer down and stuck.

One more, and then you get to do some work brainstorming your meaningful mission.

A meaningful mission thrives on empowerment. I happen to believe this statement, and it drives me: Within each person is a destiny trying to emerge. My own meaningful mission is to facilitate, *empower*, and encourage that destiny to emerge from the people I serve.

People have seen a destiny in me and for me. They have spoken empowerment over me, pulled it from me when I resisted or when I didn't believe it. They held it until I could embrace it for myself. It's my responsibility to now do that for others.

This is the basis for all great lives: *"Do unto others as you would have others do unto you"* (Luke 6:31). In my case, it's doing unto others as others have already done for me.

Imagine a person trapped in poverty, or stuck in their circumstances, felled by their choices or limited by their beliefs. They are like children living in the confines of the inner city, and for their recreation, they play in muddy puddles by the curb. This is as big as their world has ever been.

Now imagine taking that child, putting him or her on an airplane for a vacation by the ocean, moving them from mud puddles to the sea. Can you understand how expansive their world would suddenly become? Can you sense the wonder and possibility that would open in their thinking, in what they can now conceive and dream and achieve?

I exist to take people from mud puddles to oceans. It's my meaningful mission. And God has been stretching me to consider my own personal journey from small thinking to expansive thinking, from limited belief to an unlimited mindset.

I now know that for the rest of my life, I am to "Inspire, Encourage, and Lift to their Fullest God-Given Potential 100,000 People Every Week."

For me, this involves speaking and coaching. I know it involves writing and empowering. This mission has been in me all along; I just had to surrender to this high calling.

And you do, too. Your meaningful mission will be different than mine. Your mission won't necessarily look like someone else's, but here's what they will have in common:

Support

Empathy

Respect

Volunteerism

Empowerment

Now, let's roll up our sleeves and dive into crafting your meaningful mission.

As you consider how to design a meaningful mission for your life, begin with four basic, foundational questions:

1. Where am I now regarding the S.E.R.V.E. tenets of a meaningful mission?
2. Where do I want to be?
3. Why do I want a meaningful mission?
4. How do I design one?

You have to answer the first three questions honestly from your own heart. I can coach you through the *how*.

AN EXERCISE IN SELF-AWARENESS

Step One: Think about your strengths and abilities. Write down in a notebook what you believe are your greatest strengths and abilities. God buries gifts inside each of us. Our gift back to Him is to find and use them.

This is an exercise in self-awareness. You may say, "I don't have gifts." Yes, you do. You just haven't allowed yourself to think deeply enough about this. Perhaps these questions will stimulate your thinking: *What breaks your heart? What causes you to get passionate? What draws your attention every time someone brings it up? What have others said about you that they admire? What qualities have you always been glad you possessed?*

Here's a Pro-Level Self-Assessment Tip:
You are most qualified to help those who are going through what you just came through.

Let me give you an example. At age 29, Michael J. Fox was diagnosed with Parkinson's disease. You know what he's spent the bulk of his time doing since then? Helping others who have Parkinson's disease, especially in the area of research and development. His meaningful mission is the Michael J. Fox Parkinson's Foundation, which has raised $1.5 billion for research.

Are you further on the road to recovery or sobriety than someone else? There's a mission.

Did your family go through an experience that nearly wrecked you, and now you can offer help to other families facing the same devastation? There's a mission.

Did you build a successful business and now can coach others to do the same? There's a mission.

Do you have resilience and can teach others how to be resilient also? There's a mission.

For this step, write down everything that comes to mind. Spend as much time as you need fleshing that out. Your meaningful mission is already a part of your story; it's in you. You just need to take some time, look at the scope of your life, and identify some themes or categories or passions, abilities, or experiences that you have.

Here is the truth about you: You were made for a mission.

Step Two: Add your values. We learned this earlier. When you know what matters, you also know what shouldn't occupy the front of your mind or the top of your schedule in your life.

My relationship with God is my top value. This informs the kind of person I strive to be—authentic, integral, unselfish, loving, generous—that all flows from my spiritual convictions.

My wife, children, and grandchildren come next. The opportunity to serve them and others is next. That is the heart of my meaningful mission. My health follows that. Friends are also a great value to me.

See how this works? That's what you'll write down in your notebook.

Once you are clear on your values, all of a sudden, you have clarity on what you will include on your calendar. You now have clarity on how you will spend your money, on why you want to earn money, and on any new opportunities that come your way. If those opportunities don't align with your values, then they won't be added to your schedule. It's not going to be a priority.

Most people don't think intentionally about this. Most people live by default. This is about living by design.

Align your experiences, beliefs, emotions, passions, interests, abilities, talents, etc. (Step 1) with your values (Step 2). These two parts of the process alone will help you begin to see the possibility of *what* to include in your meaningful mission and *who* to include in your meaningful mission.

Step 3: Define and take action. You align everything you know to be true about yourself and everything you value. Then, with that alignment, like the tip of an arrow, you define steps you can take to begin even tentative action in the direction of your meaningful mission. This is how you start.

If even just one idea came to you, write that down, and then underneath it, write two actions you will take today. These steps might be scheduling a meeting, buying a book to get more informed, telling a significant other about your ideas, or researching who is already making an impact in the area of your interest. What you do is less important at this point than doing two things immediately.

Here are a couple of other ideas: What is a note you could send, a Venmo or PayPal you could send, a call you could make, or identify someone you will stop by to serve on your way home from work?

Write these two action steps down, and then ACT.

I've been working on my meaningful mission for a while now. This can be a long, refining process to get it robust. And you'll want to take weeks and months to keep refining yours because once you have clarity, it will guide you for the rest of your life—whatever you do in business, whatever you do in family, whatever you do forever. Your meaningful mission will be your heartbeat.

Here's what mine looks like after months of working on this: Inspire, Encourage, and Lift to Their Fullest God-given Potential 100,000 People Every Week.

I will do this as a Coach (like I do with you) by aligning my values, personality, experiences, and abilities.

I will teach value, worth, empowerment, resilience, moving past limited thinking, and overcoming fear.

I will do this from a personal foundation of:

Strong Faith

Peace of Mind

Financial Freedom

And Joy

I will consistently contribute my wisdom and experience to enrich and encourage others.

And I will continue to build out my legacy message that *anyone* can Bounce Back and Bounce Better.

Does everything have to be crystal clear to begin? No. But I have done one thing exactly right. And you can too. I started.

You just need to start.

You don't need money or even a fully formed plan. You'll make mistakes, but you will also do a ton of good along the way. You don't have to wait a single minute before starting to improve the world!

Do you see how this makes your world go from small to *huge*? You have unlimited opportunities to serve, love, lift, have empathy, and empower. This is the purpose of your life. Pick a person, pick a project, and get started.

As Karen, my friends, and I were leaving the Holocaust Museum, there was a quote on the wall that caught not only my eye but also my heart. It was written by a Holocaust survivor, and there's power in the message:

> *You who are passing by*
> *I beg you*
> *Do something*
> *Learn a dance step*
> *Something to justify your existence*
> *Something that gives you the right*
> *To be dressed in your skin, in your body hair*
> *Learn to walk and to laugh*
> *Because it would be too senseless*
> *After all*
> *For so many to have died*
> *While you live*
> *Doing nothing with your life.*
>
> –Charlotte Delbo, Auschwitz Survivor

You have been given this one and only life! Go do something great with it!

I believe in you, and I'm cheering!

CHAPTER FOURTEEN

A Custody Battle

The purpose of my life is to help you reach your fullest God-given potential. I believe there is greatness inside of you. I believe you were created for success and fulfillment and service and love. I've dedicated my life to the task of helping 100,000 each week discover and unlock their great destiny.

That said, there is another part of this conversation that is as important as anything you've read thus far. It is this singular question: Do you know who has custody of your life and future? Like parents battling for custody of a child, fighting for the right to raise and develop that child, seeking legal guardianship with the rights of decision-making, there is a battle for custody that is often waged between men and women. And it is framed in these terms: Who is responsible? Who has decision-making authority?

Too often, we want someone to swoop in and save us from being custodians of our decisions and futures. We read books (and I am thankful you are reading this one), we listen to podcasts, and we attend seminars, all in the hopes that someone, somewhere, will take responsibility for our future.

And this is where, as a personal development coach, I have to level with you: No one is coming to save you. No one is coming to be a custodian and make decisions for you.

As your coach, I can and will give you tips, strategies, best practices, and a pathway forward, *but only you can do you*. I can't do it for you, and even God won't do it for you.

My insertions of stories from my faith tradition are not meant to intimidate or even persuade you to change your personal beliefs at this time in your life. But I would not be true to myself if I didn't include them in these pages.

In the book of John (5:8 NIV), there is a story of a man who couldn't walk. When Jesus approached the man and felt compassion for the man, Jesus healed him. But then Jesus said something vital to that man and for our conversation: "Pick up your mat (a pallet-like bed) and walk."

Do you sense the personal responsibility required in that? Do you hear that there's only so far Jesus would go before the man had to take agency and responsibility and custody of his own life and future?

And you must also. This singular mindset shift is a necessary yet power-producing realization.

Taking personal responsibility for your future is the secret sauce of success.

Let's return to the primary metaphor throughout this book: When I first spoke about the ball bouncing from deflation to greater heights, I described an energy that builds within the ball. It's physics. Like a coiling snake before a strike, this energy swells and swells and then propels the ball to new levels. That energy is fueled by personal custody.

When you realize it is up to you to do something with what you've been given (like the healing and the new opportunity Jesus extended to the man who couldn't walk), you will either shrink from the responsibility and continue to lay on your mat of "not-enough-ness," imposter syndrome, fear and all the other paralyzing things that hold you back and keep you down. *Or* you will embrace the second chance, embrace the available bounce, and seize the opportunity with full custody.

At some point, after the last conference speaker finishes, after the last page of the book is read, you must decide what you will do with your one and only life.

Some find this frightening. But remember, even though you are the primary custodian of your life and future, you are not alone on the journey. A coach like me can walk beside you, though we can't walk for you. I can point the way, but you must take the steps. I can encourage your progress, but you must make that progress yourself.

I think of that man on the mat with legs that wouldn't hold him upright. And then I "see" him stand, and at that moment, he had a choice: sink back down to what he knew or take those first tentative steps in the direction of his destiny. The man in the story walked!

Why are you not doing what you desire to do? You can. Have you ever pondered the question, "I wonder what I am truly capable of?" And are you taking any steps to discover the answer?

At some point, you picked up this book for a reason. Now, it's time to walk. Destiny and greatness are in you. Your best life is waiting. Take custody of your future.

DEFINING MOMENTS

I am a big believer in defining moments: those moments in our life when we could turn this way or that, advance or retreat, learn the lesson, or shrink back only to have to learn the lesson again. What we do in the moments when two options loom before us are the moments that define us. What we do when faced with victory or defeat really do establish what we are made of and what our future can become.

In the opening pages of this book, I described one such defining moment in my life. As I lay in my bed, groggy from a drug overdose, that small voice dancing in my head was saying: *Why does God want me to live?* I could have ignored it, and I would have settled back into my sad, broken, failed life. But what I chose to do changed everything. I chose to discover *why* God wanted me to live. It was a defining moment.

The second one came in December 2021. COVID-19 raged around the world and was especially vicious in the Los Angeles area. Crematoriums asked for changes to air quality control standards because so many corpses were stacked in freezers waiting for cremation. It was a horror comparable only to the ravages of war.

I shared earlier in the book that my older brother, Jeff, became infected. He stayed at home for a few days, hoping that his case would be mild like some had experienced. After a week, it was apparent that he required hospitalization.

Jeff was my best friend for the entirety of my life. We spoke or texted every day. We joked, shared our lives, embraced, and openly expressed our love. Until that point, I had never had a day on Earth that Jeff wasn't part of. We were as close as two brothers could possibly be.

As he entered the emergency room, he confided to his wife that he didn't want to die. We all believed this large, gregarious man would pull through. It quickly became apparent, however, that his battle would be ferocious. He was put on a ventilator and into an induced coma.

None of us who loved him ever heard his voice again. We never again heard his contagious laughter. We never again felt his bear-hug embrace. After thirty days in that horrific condition, my brother, Jeff, died one day after his sixty-third birthday.

It is not overstating it to say that much of me died that day too. For months, I sleepwalked through my days, not comprehending how I could survive—or even if I wanted to—without him in this world. The crushing emotions still sweep through my heart from time to time, even after some years have passed.

But then, something penetrated my depressed mind and spirit. A thought that caught hold and took root: If my brother—who was larger than life, who seized each day and squeezed every ounce of joy and wonder from it—could no longer live a full, rich life on this Earth, then I would live it for him. I concluded the best way to honor his memory was *not to die with him but to live life as he would.* That was a defining moment.

That vision of what could be… what should be… seized me and, in a very real way, brought me back to life, but not life as I had lived it before. No. That would be a waste. I came back to life determined to see just what I could achieve. I became determined to grow, rise, explore, search, reach, climb, and discover all that I was put on this planet to experience and accomplish.

Life became a gift like it had never been before. I saw my future as I never had before.

And further, I leaned into some dreams I had tamped down for far too long. I stepped into who I truly am and stepped into the greatness my brother saw in me. My brother's death was one of the hardest experiences I have ever faced, and yet it transformed for me a far greater LIFE than I could ever have lived.

Now, my days are spent helping others like you to reach for and achieve your God-given potential; I strive to encourage, train, and equip others to stop living small and step into their own greatness and destiny.

And as I help thousands of others become who they were created to be—fully alive, fully living—I see pieces of my brother expressed in the shine from each eye of every person who has finally and fully come alive.

I want that for you. I *believe* that for you. You have greatness within you and a destiny waiting to emerge.

And just like for me after my brother's death, the way forward for you is already inside you, waiting to be released. Stop living small. Stop hiding your light for fear of what others might think. Stop stifling your dreams.

I don't believe it's an accident that you have read this book. You have within you the touch of the divine. You have a sense that you are made for more. And that fact alone should be enough to motivate you to seize every day: living more, loving more, and serving more.

You are no longer a deflated ball lying in place. Energy is building inside you. You can feel it. You are ready for more. And here's the truth: You *can Bounce Back,* and you *can Bounce Better!*

I'm living proof.

Make the rest of your life the best of your life. I believe in you, and I'm cheering!

– Brad Scot Johnson
   California, 2023

BONUS SECTION:

A Resource to Access the Key Principles Found Throughout the Book

CHAPTER ONE:

No matter who you are, what's happened to you, what you've done, how terrible it becomes, or how hopeless it seems, everyone can *Bounce Back* and *Bounce Better.*

CHAPTER TWO:

Fear, guilt, and shame were the three prominent emotions of my religious upbringing.

Your logical mind is not as powerful as your emotional mind, and it's in the emotional mind that self-doubt and "not-enough-ness" live.

CHAPTER THREE:

A *weakness* may be something you're good at, but it leaves you empty and unfulfilled. While a *strength* is also something you're good at, it leaves you satisfied and energized.

We can't heal what we don't reveal.

We are only as sick as our secrets.

You leave before you leave.

The same alignment pointed in a better direction can have proportionally the same power for good.

CHAPTER FOUR:
Waking up from that suicide attempt was the most awful, beautiful moment of my life.

"You actually see the furthest at night because in the darkness, you can see the stars."

When someone believable believes in you, you begin to believe in yourself.

Everything I need is available and coming my way, and it comes just on time. Not always my time, but on time.

I served coffee to those to whom I used to serve communion.

Forgiveness in the spiritual, quantum realm creates the perfect environment and energy for the physics of *Bounce* to exist.

Give yourself permission to believe that greatness lies within.

I must be merciful to those who are unmerciful.

You are one relationship away from your whole life changing in a positive direction.

Balls don't always bounce in predictable ways.

When you know your emotional world intimately, decisions like this become easier.

I became keenly aware of just how wonderfully orchestrated life is when we are honest seekers of direction and truth.

The resources for my next season of life were already real, present, and available. Further, it was already spiritually headed in my direction, and I had no idea.

One must seek wisdom from others, search one's own heart to assess the purity of one's motives and thinking, and then come to peace with one's own decision.

CHAPTER FIVE:

Emotions are stronger than logic.

The resources for everything you need and all you dream of are already present and moving in your direction when you are operating in belief.

CHAPTER SIX:

Critics' voices rise in direct correlation to the height of your *Bounce*.

The answer isn't coming from outside of you. It's already in you. Everything you need is in place in you. This is divine design. God created you in (with) His image.

What happens over time is that imagination gives way to logic and memory. Rather than envisioning what "could be," we settle into what is—or worse, what was.

Future = Present living allows us to experience the future now before it is reality. Remember, living from logic and memory, living in the emotions of past events is called Past = Present living, where we live the past every day in the present. Imagination allows us to live in the Future = Present.

You need to believe so deeply in the future that you both see and feel it.

An unspoken dream is an illusion.

CHAPTER SEVEN:

Starting a business or church from scratch is much like building an airplane while in flight.

Old dogs can, in fact, learn new tricks.

Keep your head, heart, feelings, words, and actions in alignment.

CHAPTER EIGHT:

The only true failure is the failure to try.

Choose your hard.

CHAPTER NINE:

The height of your *Bounce* is determined by what your mind can imagine and what you truly believe you deserve.

You cannot manifest the life you imagine if you have contradictory beliefs that block your future from becoming reality. Our achievement will only rise to the level of our beliefs being able to support that achievement.

The minute you adjust your life to fit the opinions of others, you have given up your power to create the life you want. And you have given power over your life to the very people who are haters.

CHAPTER TEN:

Disappointment is unmet expectations.

You may not have the life you thought, but you can have the life you dream.

Pivot: Adapt to change, tap new markets, and capitalize on new opportunities.

Your future success in life is way more about the hunger in you and way less about the history behind you!

Those who pivot, persevere.

"…my core business was larger than one product or method."

This is the step where you imagine the future as it could be, not as it is.

There are 1,440 minutes in a day. Just 20 of those can radically change how the entirety of your day goes.

Life doesn't have to be the way you thought it would be in order to have a great life.

CHAPTER ELEVEN:

"People-pleasing": Compromising or abandoning who we are, what we want, or what we value in order to be validated, included, or loved.

If it isn't working for me, I can change!

Show up for yourself!

The world needs my fully authentic, full light!

CHAPTER TWELVE:

Everyone has his or her own journey.

If you could do something with your life and you knew you wouldn't fail, what would you do?

Life has a way of shrinking your dream to match your faith. Let's be honest: life has a way of beating confidence and courage out of you.

Fear diminishes who you are and what you can achieve.

Behind the fear of judgment is the fear that we are not enough.

Everything you need is already in you.

God's spirit is in you.

Everything you ever wanted is on the other side of fear.

Mental breathing is visualizing a positive outcome for the area where you are currently afraid.

An unsure step forward is still a step forward.

The main goal of playing not to lose is running out the clock.

CHAPTER THIRTEEN:

Support

Empathy

Respect

Volunteerism

Empowerment

Here's a Pro-Level Self-Assessment Tip: You are most qualified to help those who are going through what you just came through.

CHAPTER FOURTEEN:

Taking personal responsibility for your future is the secret sauce of success.

THANK YOU for reading my book! To express my gratitude, I would like to give you a free BONUS!

SCAN THE QR CODE:

I appreciate your interest in my book and value your feedback as it helps me improve future versions of this book. I would appreciate it if you could leave your invaluable review on Amazon.com with your feedback. Thank you!

Made in United States
North Haven, CT
24 September 2024